GENETIC RECOMBINATION

Thinking About It in Phage and Fungi

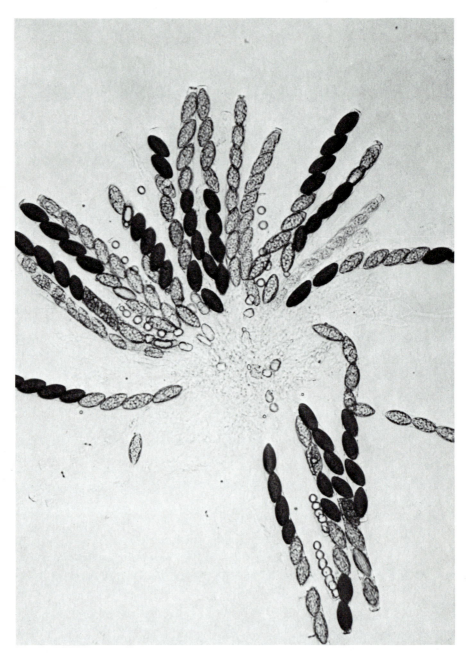

Neurospora asci illustrating first and second division segregation at a spore-color locus. Second division segregation signals recombination between the spore-color gene and its centromere. (Photo courtesy of D. R. Stadler.)

GENETIC
RECOMBINATION

Thinking About It in Phage and Fungi

Franklin W. Stahl
UNIVERSITY OF OREGON

W. H. Freeman and Company
SAN FRANCISCO

A Series of Books in Biology

Cedric I. Davern, *Editor*

Cover photograph of a "Holliday Structure" courtesy of H. Potter and D. Dressler.

Sponsoring Editor: Arthur C. Bartlett
Project Editor: Nancy Flight
Manuscript Editor: Howard Beckman
Designer: Perry Smith
Production Coordinator: Fran Mitchell
Illustration Coordinator: Batyah Janowski
Artist: Eric G. Hieber, E H Services
Compositor: Syntax International
Printer and Binder: The Maple-Vail Book Manufacturing Group

Library of Congress Cataloging in Publication Data

Stahl, Franklin W
  Genetic recombination.

  (A Series of books in biology)
  Includes index.
    1. Genetic recombination.   2. Bacteriophage—
Genetics. 3. Fungi—Genetics. I. Title.
QH448.S7   575.1   79-13378
ISBN 0-7167-1037-4

1 2 3 4 5 6 7 8 9

To Eleanor and Oscar, who encouraged me
to try, and to Mary, who kept the experiments
going while I did.

Contents

Preface

Genetic recombination is ubiquitous; all forms of life engage in either occasional or periodic reshuffling of the cards in their genetic decks. In the Darwinian view, ubiquity implies adaptive significance of the highest order, and to a biologist that signifies importance. We must study genetic recombination because it is important. But there are other good reasons for our study. Recombination in our service is useful. With it we can construct genotypes for agricultural, horticultural, pharmaceutical, or investigational use. A better understanding of recombination will thus lead to its more effective use. Nevertheless, these reasons for studying recombination sound a bit more like excuses than reasons to me. The students of recombination whom I know pursue the study because it is an amusing, baffling, and well-defined but surprisingly slippery intellectual challenge. If that is at least part of your motive for reading this book, I think we will get along.

I have organized the subject in the following way. I recognize that a molecular understanding of recombination is our goal, and the final chapters will speculatively wed our knowledge of molecules with our knowledge of recombination. The first nine chapters set forth that knowledge. That the first eight are "genetical" and only the ninth "chemical" reflects primarily an acceptance of my own limitations; I thought it better to develop the subject from relative strength than to sacrifice fidelity to balance.

The eight "genetical" chapters begin with elementary concepts established early in the history of genetics. Complexities are introduced in a sequence chosen for heuristic reasons, and though I have distorted history shamelessly, I have never done so gratuitously.

Phage and fungi have played leading roles in our present understanding of recombination. Fungi have been important because they combine the attributes of microbes with the orderliness of the recombinational stage of the life cycle in eukaryotes. Phage have been helpful because their small chromosomes can be studied physically as well as genetically. Over the last twenty years, ideas have reverberated between the viral and fungal schools, and this intercourse has been institutionalized in a remarkable series of biennial workshops sponsored by the European Molecular Biology Organization (EMBO).

At what level can this book be used with profit? I think upper division undergraduates in genetics would find it suitable. Graduate students in biology or biochemistry should be up to it, especially if they enjoy the kind of logic that underlies genetic arguments. In fact, I hope the book lays bare the logic of current experiments and arguments in recombination to anyone with some background in genetics and a willingness to work.

Although you will encounter equations and derivations, those who like mathematical statements will find my math to be child's play. Those who get headaches from integral signs will find that the arguments are laid out in a way that permits the reader to say "thus and so" at each equation without fatal loss of continuity. Some elementary probability notions are absolutely unavoidable—that's how genetics is—but anyone who has survived a first course in genetics should be able to manage the notions used here. In the appendix, which should be studied only after Chapter 6, I go on a mathematical binge; skip that if you wish.

Definitions have been a problem, as they are in any field in which changing concepts impose changes in meaning on words originally well defined. In a few instances I have found it necessary to invent new words; in most instances I have taken care to define the way in which I shall use old ones. When I have slipped, the context will generally prevent misunderstanding.

Problems are provided at the end of each chapter. They serve several purposes, and that may cause trouble. Some of the problems are the sort that a student expects to find in a textbook—they check the reader's understanding of the chapter or provide exercise in relevant basic reasoning. Other problems treat ideas that are a bit too mathematical or parenthetical to include in the main body of the chapter. Yet others invite readers to explore ideas beyond the general level of the book. Read all the problems, but use your own taste and good sense in choosing ones to solve. In any case, solutions to most of the problems are given, so feel free to peek if that will help you decide whether a given problem is worth your effort.

I have already implied the debt that this book owes to EMBO and to Robin Holliday and Neville Symonds, organizers of the workshops. The Genetics Department of the Hebrew University, Jerusalem, provided hospitality, help, and an inspirational atmosphere that allowed me to get the project underway. The Lady Davis Foundation brought us to Jerusalem, and the John Simon Guggenheim Memorial Foundation supported us while we were there. Original experiments reported here as well as the scientific ambience within which the ideas have grown have been supported by grants from the National Science Foundation and the National Institutes of Health.

A preprint version of this book was tested on well-warned University of Oregon students, staff, and visitors in Biology 487, Spring Quarter, 1977. Their comments were helpful in the preparation of the final manuscript. Absolutely invaluable were the criticisms of Gary Gussin, Jeff Haemer, David Stadler, Ric Davern, and Dana Carroll. I realize now that I failed to list one really good reason for studying recombination—recombination geneticists are fine company. I am grateful for the help they have extended to me in my efforts to understand recombination and to present that understanding to you.

FRANKLIN W. STAHL
June 1979

GENETIC RECOMBINATION

Thinking About It in Phage and Fungi

1

Random Spore Analysis

Genetic recombination results from exchanges between homologous chromosomes. Two marked genes recombine when an odd number of exchanges occurs in the interval between them. The probability of an odd number of exchanges increases with increasing distance between genes. As a result, recombination frequencies reflect distance, allowing linkage maps to be constructed. In most eukaryotes, fungi included, exchanges are not independently disposed along a chromosome; instead, one exchange interferes with the occurrence of others nearby.

THE GENETIC CROSS

Somewhere in the life cycle of all sexually reproducing eukaryotes, two haploid nuclei fuse, producing a diploid nucleus (Figure 1-1).* Following this event, the diploid cells undergo reduction, restoring the haploid con-

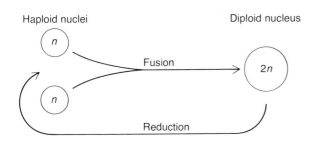

FIGURE 1-1

A schematized life cycle of a fungus. In Fungi, as in other sexual eukaryotes, haploid nuclei are fused from genetically different clones. During meiosis the diploid cells (or some of their descendents) accomplish a reduction in chromosome number to restore the haploid state. Meiosis also provides the occasion for genetic recombination. The haploid number of chromosomes in organisms is customarily symbolized by n.

---

* The glossary defines terms that may be unfamiliar.

dition. In this basic life process genotypes are often altered by processes of recombination. These processes can be studied by conducting a cross (fusion) of hereditarily distinct haploid cells and determining the genotypes of the subsequent products of reduction. The details of the methodology depend of course on the manner in which the organism being studied has varied the theme of this basic process. The analysis is relatively direct in fungi, with their free-living haplophase; in general, the genotype of a haploid cell product can be directly inferred from the phenotype of the clone to which it gives rise (but we shall encounter important exceptions to this general rule). The operations of a cross for two widely studied fungi are shown in Figure 1-2.

Genetic recombination is observed when the two gametes differ by at least two mutations. When such is the case, the genotypes of the pooled haploid products of a number of acts of reduction will be more various than the original parental genotypes. Thus, the twin process of fusion and reduction promotes the formation of new combinations of hereditary factors. The processes whereby these new combinations are created are the subject of this book.

The kinds of factors available for use in fungal crosses are varied. Among the most useful have been auxotrophy, drug resistance, sugar-fermentation ability, and pigmentation. Suppose we cross a haploid culture of a mutant requiring a single amino acid with a culture of a mutant resistant to a drug. We then examine a sample of the haploid products from a large number of reductions for the frequency of the possible types: auxotroph-sensitives and prototroph-resistants, like the parents, and prototroph-sensitives (wild type) and auxotroph-resistants (double mutants), the new types. We can talk of these possibilities more efficiently by using symbols. So, let a be the symbol for the factor determining the mutant character auxotrophy, and let a^+ stand for its wild-type allele determining prototrophy. Similarly, let b indicate the factor for drug resistance and b^+ the wild-type allele determining sensitivity. We shall refer to the gene having to do with the auxotrophy in question, without regard to its allelic state (a or a^+), as the A gene; similarly, we shall refer to the B gene. We can then diagram our cross as follows:

$$ \begin{array}{c} ab^+ \\ \times \\ a^+b \end{array} \quad \text{fusion} \longrightarrow \begin{array}{c} ab^+ \\ a^+b \end{array} \quad \text{reduction} \left\langle \begin{array}{l} \text{Parental} \\ \text{types } (ab^+, a^+b) \\ \\ \text{Recombinants } (a^+b^+, ab) \end{array} \right. $$

The primary data here are the frequencies of the four types of haploid reduction products. These four frequencies, however, can be summarized by a single number, thanks to the following rule:

RULE 1. Both parental types are statistically equal in frequency and the two recombinant types are statistically equal in frequency.

The results of a cross are therefore expressed as R, the recombination frequency, which is the number of recombinants divided by the total number of haploid cells in the sample examined.

GENETIC MARKERS

Another rule of recombination studies is the following:

RULE 2. The recombination frequency depends only on the mutants involved and not on the combinations that define the parental types.

Thus, the cross $a^+b^+ \times ab$ yields the same recombination frequency (the number of ab^+ and a^+b divided by the total haploid cells examined) as that in the cross $ab^+ \times a^+b$ (the number of ab and a^+b^+ divided by the total haploid cells examined).

The widespread applicability of Rules 1 and 2 is fundamental to our thinking about recombination. They provide partial justification for the concept of a *genetic marker*. Genetic markers are mutations used to reveal recombination without influencing it. Throughout Chapters 1–4 we shall deal only with crosses in which the mutations employed behave as good markers, i.e., crosses that adhere to Rules 1 and 2. In subsequent chapters we shall see that the marker concept is an idealization; crosses using some mutations *do* depart from Rules 1 and/or 2. These departures, although painful, are nevertheless of great interest for what they promise to tell us about recombination processes.

GENETIC LINKAGE MAPS

In the reduction process of meiosis, paired chromosomes are independently oriented (Chapter 3). Thus, gene pairs on different (nonhomologous) chromosomes have recombination frequencies equal to one-half. The random assortment of chromosomes during meiosis is the most important mechanism of genetic recombination in eukaryotes because it is the mechanism by which most pairs of genes recombine. Nevertheless, it will not concern us in this book. As far as we know, the individually visible chromosomes are distinct

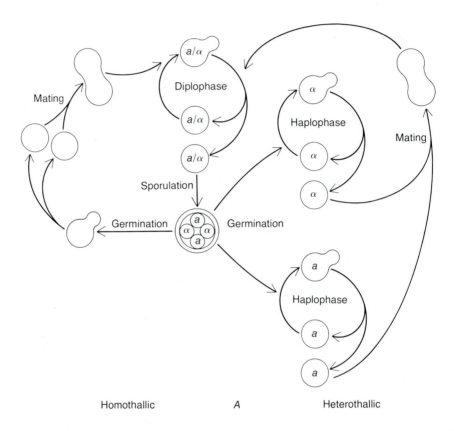

FIGURE 1-2

Operations involved in a genetic cross in two fungi. (**A**) Life cycles of two *Saccharomyces* strains. Two haploids of mating type *a* and two of mating type α comprise a four-spored ascus, which is formed after sporulation of a heterozygous diploid *a*/α. The stable states of heterothallic strains are diploid and haploid, while homothallic strains have only one stable state, the diploid, since the transient haploid cells rapidly diploidize. Crosses involve matings then sporulation. Matings are conducted by juxtaposing (generally on an agar surface) *a* cells of one genotype and α of another. Diploid products of mating (zygotes) can be recognized by their irregular shape, or they can be selected on a growth medium that supports neither of the two complementing haploid types. A single isolated diploid cell will give rise to a culture that can be induced to sporulate by nitrogen starvation. Genotypes can be analyzed by sampling the resulting haploid cells pooled from many asci after the spores germinate, or alternatively, by dissecting individual asci before they burst and analyzing their contents.

(**B**) Vegetative cultures of *Neurospora* consist of branched, segmented filaments (hyphae) with each segment containing many (usually identical) haploid nuclei. Hyphal tips pinch off to form vegetative spores (conidiospores) containing one (micro-) or several (macro-) nuclei. When vegetative cultures of opposite mating types (*A* and *a*) are grown in proximity, each responds by

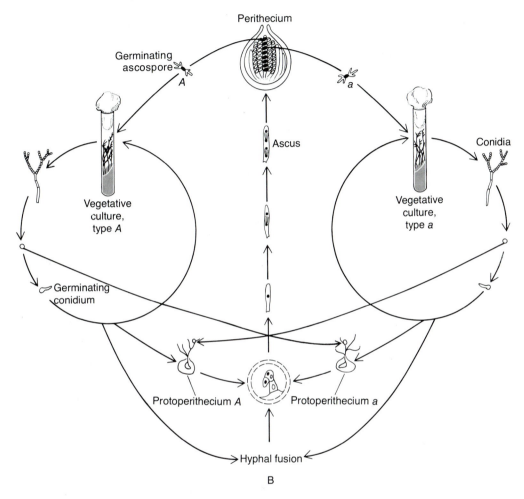

Peritechium

Germinating
ascospore

A

Ascus

a

Conidia

Vegetative
culture,
type *A*

Vegetative
culture,
type *a*

Germinating
conidium

Protoperithecium *A*   Protoperithecium *a*

Hyphal fusion

B

forming protoperithecia, organs specialized for receiving conidiospores of the opposite
mating type. Cell fusion then gives rise to a dikaryotic cell whose repeated divisions result in
a mass of ascogenous hyphae in the perithecium. Nuclei in each dikaryotic cell then fuse to
form the diploid zygote, which promptly undergoes meiosis within an elongated ascus. The
products of the meiotic divisions remain aligned so that the two nuclei at each end of the
ascus are sisters of the second division. A postmeiotic mitosis precedes the formation of
spore walls so that each mature ascus contains eight aligned haploid cells. Germination of an
ascospore gives rise to a new vegetative hyphal mass. Crosses are conducted by sprinkling
conidiospores from one mating type on protoperithecia of the other mating type. A sample of
the resulting ascospores can then be analyzed for recombination frequencies. Or, individual
asci can be dissected prior to spontaneous spore liberation and the genotype of each spore
determined.

physical entities, making their random assortment inevitable. This is not so, of course, for genes on the same chromosome; their recombination involves interactions between homologous chromosomes, and it is the nature of those interactions that we aim to elucidate.

Genes on the same chromosome have a recombination frequency less than or equal to one-half. We say that such genes are *linked* with each other. Different pairs of genes show different degrees of linkage; some have recombination frequencies less than 1 percent, while others have recombination frequencies indistinguishable from 50 percent. A set of genes whose members are linked with one another is called a *linkage group*.

The results of crosses involving different pairs of markers in the same linkage group can be summarized in the form of a genetic linkage map.[1]* The following rules guide the construction of such a map. (1) A line segment is drawn. (2) Symbols for the two genes giving the largest recombination frequency with each other are placed at opposite ends of the segment. (3) Symbols for other genes are located between these two terminal ones in such a way that genes giving low recombination frequencies with each other are located closer together than those giving larger ones. In the absence of complications, this simple procedure permits the placing of the gene symbols in a unique order on the line segment. The result is a linkage map.

The mappability of genes implies the operation of yet another basic rule of recombination studies.

> RULE 3. The recombination frequency is a function not of the markers themselves nor of the genes within which they lie but of the remoteness of the markers from each other on the genetic linkage map.

Allow me to illustrate this somewhat tautologous Rule of Mappability. When we say a map order is *ABCD*, we imply that *A* and *D* show the largest recombination frequency (*R*), that *R* for *B* and *D* is greater than that for either *B* and *C* or *C* and *D*, and that *R* for *A* and *C* is greater than that for either *A* and *B* or *B* and *C*. Now consider the case where *B* and *C* have only slightly different recombination frequencies in crosses with *D*. Then these two genes also have only slightly different recombination frequencies in crosses with *A* and with each other. The operation of Rule 3 demands an additional property of a genetic marker. A proper marker can be given a map position consistent with all the measured recombination frequencies.

The above three rules are the cornerstone of recombination analysis. They guided studies until the 1960s, when nontrivial exceptions became

---

* All numbered notes appear at the end of the chapter.

apparent in crosses involving very close markers.[2] We now recognize that these three rules are approximations whose validity is limited but whose usefulness is undeniable. In Chapter 6 we shall examine the results of crosses between *very close* markers and struggle with the uncertainties that characterize such results.

Genetic linkage maps for eukaryotes have two additional features. For three genes linked in the order ABC, $0.5 < R_{AC}/(R_{AB} + R_{BC}) < 1$. It is observed that (1) the ratio approaches unity for small R values (as long as R is not so small that Rules 1–3 break down), and (2) (the ratio approaches 0.5 as R_{AB} and R_{BC} each approach 0.5. Property 1 is the law of additivity of small recombination frequencies. Property 2 implies that recombination frequencies approach 0.5 as markers get further and further apart on the map.

So far in our discussion of genetic linkage maps I have avoided any mention of the order of genes *on a chromosome*. I have done so in order to underscore the definition of a genetic linkage map as a summary of recombination frequencies observed in a set of genetic crosses involving linked markers. The demonstration that the order of genes on a genetic linkage map corresponds with the arrangement of the genes on their chromosome was a triumph of classical genetics.[3] This correspondence of map and chromosome provides a model for explaining the mappability of markers. The model I shall describe below is both the simplest imaginable and the earliest historically. Our progress through the book will be marked by the successive refinements to which we submit this primitive model.

THE MODEL MARK I

We assume that recombination is a result of the formation in a diploid cell of a chromosome that derives parts of its length from the chromosome of one parent and other part(s) from the paired homologous chromosome of the other parent. A point where parts are joined is called an *exchange*. An exchange can occur at any of a large number of points along the chromosome, but in any single act of meiosis it occurs at only a few points. Recombination is observed when an exchange occurs at a point between two or more marked genes. For example, in the exchange diagrammed below, A and B and A and C are recombined but not B and C.

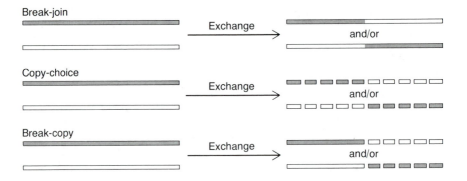

Break-join

Exchange → and/or

Copy-choice

Exchange → and/or

Break-copy

Exchange → and/or

FIGURE 1-3
A number of schemes for the production of recombinant chromosomes can be (and have been) proposed. The three illustrated here are simple *formal* possibilities adequate for thinking about the facts and concepts of Chapters 1 and 2. In the break-join scheme, parental chromosomes break and the parts are united in new combinations. In the copy-choice scheme, parental chromosomes direct the synthesis of *new* chromosomes (*broken lines*) in a cooperative way. The break-copy scheme is a bastard of the other two.

At this stage we are mute with respect to a large number of questions that could be asked about exchange. In particular, we need not now specify whether recombinant chromosomes inherit atoms as well as information from the parental chromosomes, nor need we specify whether or not complementary recombinants (ab^+c^+ and a^+bc in the above diagram) arise simultaneously as a result of an exchange. Nevertheless, we must insist that complementary recombinants arise at equal rates to insure the operation of Rule 1 (Figure 1-3). Since distant genes are more apt to have an exchange somewhere between them than are close genes, recombination frequency will, as demanded by the facts, be a function of distance.

Our model is further specified by the results of the cross diagrammed above. Among the recombinants a^+b and ab^+, some are observed to be recombinant for genes B and C as well; i.e., some of the products of reduction are ab^+c and a^+bc^+. Thus, some chromosomes will have undergone more than one exchange, as in the following:

In this example the occurrence of two exchanges between genes A and C does not lead to their recombination (with each other). In general, any even number of exchanges between two markers will leave them in parental combination; only odd numbers of exchanges will recombine them.

We can complete the Model Mark I by specifying one more feature. Exchanges must be distributed along chromosomes according to a (very lenient) set of rules. (1) For close markers, a single exchange is more probable than any larger number. (we shall hedge on this point in Chapter 6). (2) For distant markers, the probability of an odd number of exchanges cannot exceed $\frac{1}{2}$; this ensures that the maximum R value observed for linked markers will not be greater than 0.5.

THE HALDANE FUNCTION

It will be useful to us now, as well as later, to examine a *specific* simple rule that provides for additive R at small R and for R_{max} not greater than 0.5. This rule supposes that exchanges on any specified segment of a chromosome are randomly (Poisson) distributed among the haploid products of reduction with a mean number that defines the length of the segment (Figure 1-4).* For a Poisson distribution of mean x, the probability of an odd number is

$$P_{odd} = \tfrac{1}{2}(1 - e^{-2x}) \qquad (1\text{-}1)$$

Since recombination only occurs with an odd number of exchanges, we can use this same equation for the recombination frequency:

$$R = \tfrac{1}{2}(1 - e^{-2x}) \qquad (1\text{-}2)$$

Equation 1-2 is graphed in Figure 1-5. Note that R is equal to x when x is small. Thus, the assumption of randomness accounts for the additivity of small R values. Note, too, that when x is large, R is independent of x and has a value of $\frac{1}{2}$. Thus, the model accounts for the observed asymptotic approach of R to a limit of 0.5 for distant markers. Equation 1-2 relates the frequency of recombination (R) to map distance (x). It is the first of several mapping functions that we shall encounter. We call it the Haldane Function after the geneticist who first wrote it down.[4] The simplicity of the Haldane Function, with its assumption of random (Poisson) exchange distribution,

---

* Note that this model, and others similar to it, measures the length of chromosome segments in terms of the mean number of *exchanges*. It is therefore noncommittal regarding any assumptions about the uniformity of the mean number of exchanges per unit of *physical* length. The intensity of exchanges per unit of physical length could even change along the chromosome and our rule would not care. If *we* care, and in later chapters we shall, we must do experiments other than crosses in order to get the needed information to compare genetic length (mean number of exchanges) with physical length (numbers of base pairs).

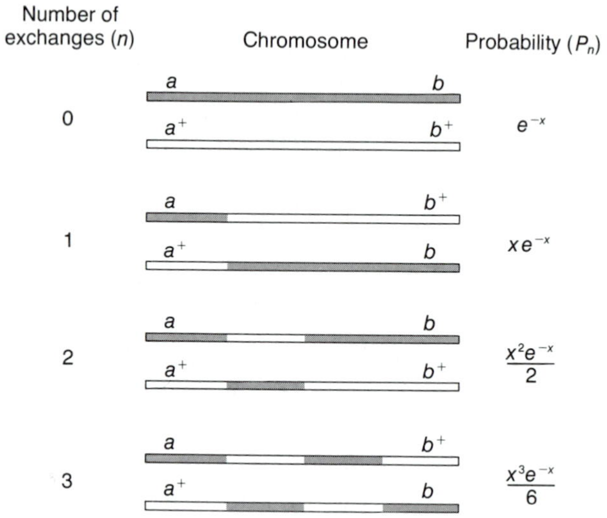

Number of
exchanges (n)   Chromosome   Probability (P_n)

a		b
0		
a^+		b^+	e^{-x}

a		b^+
1		
a^+		b	xe^{-x}

a		b
2		
a^+		b^+	$\dfrac{x^2e^{-x}}{2}$

a		b^+
3		
a^+		b	$\dfrac{x^3e^{-x}}{6}$

FIGURE 1-4

Poisson distribution of exchanges between chromosomes within a segment for which the mean number of exchanges is x (this mean is used to define the length of the segment). Poisson's distribution states that the probability of n when the mean is x is $P_n = \dfrac{x^n e^{-x}}{n!}$.

Thus some chromosomes have no exchanges, some have one, some two, and so on.

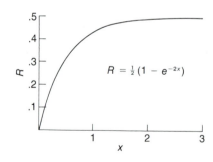

$$R = \tfrac{1}{2}(1 - e^{-2x})$$

FIGURE 1-5

The relation between recombination frequency (R) and genetic linkage distance (x) according to the Model Mark I operating under the assumption of a Poisson distribution of exchanges among chromosome segments of the same *genetic* length (see Figure 1-4).

recommends it as a conceptual landmark to which we will refer as we delve further into the realities of mapping data.

The assumption of a random exchange distribution implicit in Equation 1-2 can be tested. Multifactor (usually three-factor) crosses provide a more sensitive test, but it is instructive to look first at a test based on data from two-factor crosses.

THE ADDITIVITY TEST

If a particular mapping function is a good description of nature, then the function will transform a set of R values into x values that are additive. We can illustrate this additivity test using the Haldane Function. Suppose the following R values were observed in crosses involving the linked genes A, B, and C:

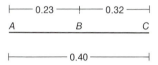

Note that these R values are *not* additive, that is, $R_{AC}/(R_{AB} + R_{BC}) = 0.40/0.55 < 1$. However, using equation 1-2 or its graph in Figure 1-5, we can convert these R values to x values, as follows:

```
       ├──── 0.30 ────┼──── 0.50 ───┤
   A              B               C

       ├─────────── 0.80 ──────────┤
```

Since I selected the R values to fit the Haldane Function, the derived x values are found to be additive. Note that a result of our test is a little genetic linkage map in which the intervals between genes are *distances*, which, like the distances on a road map, are additive. The map is *rectified*. Later in this chapter we shall resume the discussion of techniques for the construction of such rectified maps.

The additivity test can be extended. If the Haldane Function accurately describes a set of R data, then the sum of the x values for any number of adjacent values must equal the x value for the inclusive interval. For example, the following R values

```
   ├── 0.23 ──┼── 0.32 ──┼── 0.28 ──┤
  A         B          C          D

   ├──────────── 0.45 ────────────┤
```

can be converted to the following x values:

```
   ├── 0.30 ──┼── 0.50 ──┼── 0.40 ──┤
  A         B          C          D

   ├──────────── 1.20 ────────────┤
```

Note that, as a result of my foresight, $x_{AB} + x_{BC} + x_{CD} = x_{AD}$.

THREE-FACTOR CROSSES AND THE COEFFICIENT OF COINCIDENCE

The Haldane Function can be tested by the results of three-factor crosses. The cross $abc \times a^+b^+c^+$ yields the following eight types of progeny:

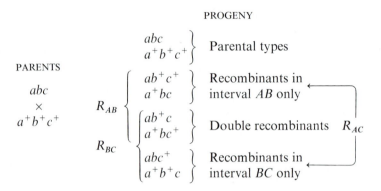

PROGENY

PARENTS

abc
$\times$
$a^+b^+c^+$

abc
$a^+b^+c^+$ } Parental types

R_{AB}
ab^+c^+
a^+bc } Recombinants in interval AB only

ab^+c
a^+bc^+ } Double recombinants R_{AC}

R_{BC}
abc^+
a^+b^+c } Recombinants in interval BC only

If exchanges are randomly distributed, then recombination in the intervals AB and BC will be independent events, and the probability of simultaneous recombination in the two intervals will equal the product $R_{AB}R_{BC}$. Deviations from independence are expressed as the coefficient of coincidence, S, as follows:

$$S = \frac{\text{observed frequency of double recombinants}}{R_{AB}R_{BC}} = \frac{R_{\text{doubles}}}{R_{AB}R_{BC}} \quad (1\text{-}3)$$

When $S = 1$, randomness is implied.

Note that S can be estimated (though with reduced sensitivity) from two-factor crosses, a point we shall make use of later. In the table of genotypes above, we see that

$$R_{AC} = R_{AB} + R_{BC} - 2R_{\text{doubles}} \quad (1\text{-}4)$$

Combining Equations 1-3 and 1-4 gives

$$R_{AC} = R_{AB} + R_{BC} - 2SR_{AB}R_{BC} \quad (1\text{-}5)$$

and, by rewriting,

$$S = \frac{R_{AB} + R_{BC} - R_{AC}}{2R_{AB}R_{BC}} \quad (1\text{-}6)$$

In Equation 1-6 we see that all the information for determining S is available from the results of two-factor crosses. However, an estimate of S from two-factor crosses is based not on a directly measured frequency, as it is in a three-factor cross, but instead on the difference between two numbers of comparable magnitude measured in separate crosses, i.e., on $(R_{AB} + R_{BC}) - R_{AC}$. Despite this practical shortcoming we shall have occasion to resort to the two-factor method. For the moment it is instructive to note that the observation of (perfect) additivity of R values ($R_{AB} + R_{BC} = R_{AC}$) implies $S = 0$, i.e., a complete absence of double recombinants in the interval AC. Later in this chapter we shall make use of the interrelations between the two- and three-factor tests for the randomness of exchange distribution. For now we note that most eukaryotes do not behave according to the concept of randomness. The nature of the departures from randomness can be stated more fully after we examine the reduction process (meiosis) in Chapter 3. Nevertheless, we can state here that the departures manifest themselves as S values less than unity when R values are rather small, implying that exchanges have a kind of territoriality; the presence of one exchange lowers the probability of finding another exchange nearby. One cannot help but feel that a clue to a mechanism of recombination lies in this observed departure from randomness, and we shall speculate about that in later chapters.

MORE ON MAPPING FUNCTIONS

When the Haldane Function proves inadequate, which it does almost everywhere, how shall we proceed in the construction of rectified maps? Several avenues are open.

FUNCTIONS BY INSPIRATION

When the Haldane Function fails, we can dream up one of our own. This amounts to writing a function that *might* work, using it to convert observed R to x, and then testing the x values for additivity. Our aim is achieved when the resulting x values show no systematic departures from additivity. Several functions have been written that adequately convert R to additive x for one or another eukaryote.[5] The graph of a function dreamed up for this occasion is shown in the solution to Problem 1-10 (Figure S-1). It has properties that characterize recombination in most eukaryotes. Its R values are more nearly additive than are those of the Haldane Function (i.e., it shows interference), and R approaches 0.5 as x gets large.

FUNCTIONS BY BRUTE FORCE

If many markers have been ordered, the smaller intervals will manifest approximately additive R values. A map can be constructed with such data by equating distance with those small values. Then the distance for markers not so close together is simply equal to the sum of all the elementary intervals between them. Should any elementary interval be a bit too long to be additive, it can be converted to distance by comparing it to an interval of (about) the same R that embraces a number of intervals that are small enough to be additive. The approximate additivity of eukaryotic R values over a wide range, a consequence of interference, facilitates this approach.

FUNCTIONS BY THE BOOT STRAP

This method makes use of data from three-factor crosses.[6] The data must be collected from a set of crosses involving a wide range of R values, but it is not necessary that any part of the map be densely populated with markers. First, S values are determined from the three-factor crosses. In a cross of the type $abc \times a^+b^+c^+$ the intervals AB and BC should be about equal, and we shall imagine them to be precisely so. The set of crosses provides measures of S as a function of R; these can be graphed and a curve drawn through them. For most eukaryotes such a curve of S versus R would show S close to zero at small R and rising to unity as R approaches 0.5. The graph is then used in the construction of the mapping function. We select an R value (R_1) so small that it is essentially linearly dependent on x. We then use this R value as our elementary interval, measuring larger intervals as multiples of it. The interval containing two elementary intervals will have an R value of

$$R_2 = 2R_1 - 2SR_1{}^2$$

We solve for R_2 by substituting R_1 and the appropriate value of S read from the graph. The R value from an interval made of four elementary intervals is then found from

$$R_4 = 2R_2 - 2SR_2{}^2$$

by substituting the R_2 obtained above and the S value from the graph corresponding to R_2. Multiples, R_8, R_{16}, etc., are found by reiterations of this process. The obtained set of calculated R values is then graphed with R_2 versus $2R_1$, R_4 versus $4R_1$, R_8 versus $8R_1$, etc., and the points are joined by a smooth curve. You can practice this method in Problem 1-9.

Whatever method we apply for rectifying, the result is a map in which the distances between loci represent the mean number of exchanges between those loci. If a *physical* map can be constructed, then it is practical to ask whether or not the exchanges are uniformly probable along the chromosome. If they are, then the linkage map is a scale drawing of the chromosome itself. If they are not, then we may inquire into the factors influencing the local exchange intensity, with the hope of learning something about the chemistry of exchange.

PROBLEMS

1-1   Two haploid cells whose genotypes were a^+b and ab^+, respectively, united to form a diploid. A culture was grown up from this diploid and cells in it were then encouraged to undergo reduction. The resulting haploid cells were of four genotypes. (a) What were they? Which are the parental types and which the recombinant types? (b) Which of the genotypes occurred in equal frequencies to each other? (c) Which genotypes were less than or equal to 25 percent in frequency?

1-2   Two haploid cells whose genotypes were a^+b^+ and ab, respectively, united to form a diploid. A culture was grown up from this diploid, and cells in it were encouraged to undergo reduction. Among the resulting haploid cells, the frequency of a^+b^+ was 0.30. (a) What were the frequencies of the other three types? (b) Suppose the cross were conducted starting with genotypes ab^+ and a^+b. What would be the frequency of each of the four kinds of haploid products of the cross?

1-3   (a) For the cross $a^+b^+ \times ab$, $R = 0.24$, and for the cross $a^+d \times ad^+$, $R = 0.16$. What can be said about the order of the genes A, B, and D? (b) For the cross $b^+d^+ \times bd$, $R = 0.38$. What can be said about the order of genes A, B, and D?

1-4   Calculate the coefficient of coincidence in Problem 1-3.

1-5   This problem involves a "marker effect." However, the example chosen is fully understandable in terms of the concepts presented in Chapter 1. (It involves a violation of our contention that $R_{AC}/(R_{AB} + R_{BC}) < 1$.) For the crosses

$$a^+b^+ \times ab, \quad R_{AB} = 0.1$$
$$b^+c^+ \times bc, \quad R_{BC} = 0.1$$
$$a^+c^+ \times ac, \quad R_{AC} = 0.25$$

(a) Calculate the apparent coefficient of coincidence using Equation 1-6. Why is the result a negative number?

(b) Now suppose you did a three-factor cross involving the same "markers" and found $R_{AB} = 0.1$; $R_{BC} = 0.1$, and the frequency of double recombinants $= 0.005$. Using these data, calculate S. What value for R_{AC} was found in this cross? How can it be different from that found in (a)?

(c) Suppose you performed the cross $a^+bc^+ \times abc$, and found R_{AC} to be the same as in (b). You then have the facts necessary to support an explanation of the "marker effect" of b strictly in terms of our theorem that R is a function of distance. What is your explanation?

1-6 Three genes, A, B, and C, are known to be linked, but their map order remains to be determined. A three-factor cross was performed, and the following frequencies of haploid progeny were noted:

$$
\left.\begin{array}{l} abc \\ a^+b^+c^+ \end{array}\right\} \quad 0.09
$$

$$
\left.\begin{array}{l} a^+bc \\ ab^+c^+ \end{array}\right\} \quad 0.19
$$

$$
\left.\begin{array}{l} abc^+ \\ a^+b^+c \end{array}\right\} \quad 0.01
$$

$$
\left.\begin{array}{l} ab^+c \\ a^+bc^+ \end{array}\right\} \quad 0.71
$$

(a) What were the genotypes of the two parents?

(b) What are the three recombination frequencies (R_{AB}, R_{BC}, and R_{AC})?

(c) What is the map order?

(d) Which genotypes are double recombinants?

(e) What is the value of the coefficient of coincidence, S?

Note that the double recombinant is the least frequent of the genotypes. Appreciation of this permits determination of the gene order without first calculating the three R values. When S is less than 1 or when the R values are small, the double recombinant class is far smaller than any of the others, making three-factor crosses a much more reliable method of determining map order than a comparison of the R values from three separate two-factor crosses.

1-7 The data below are a concocted set of R values for a creature in which $S = 1$ at all values of R. (a) What is the order of the genes? (b) Construct a rectified map from the following data:

Interval	R
AB	0.35
AC	0.16
AD	0.40
BC	0.28
BD	0.47
CD	0.43

1-8 The data below are a concocted set of R values for a creature whose exchanges show interference. In this problem the order of genes is alphabetical. Using "brute force," construct a graphic mapping function and a rectified map.

Interval	R	Interval	R
AB	0.025	EF	0.192
BC	0.025	AF	0.344
AC	0.050	FG	0.344
CD	0.050	AG	0.480
AD	0.099	GH	0.480
DE	0.099	AH	0.500
AE	0.192		

1-9 In the set of concocted data below R_1 and R_2 are recombination frequencies for adjacent intervals of about the same length. For each such pair of intervals the sum $R_1 + R_2$ is given along with the observed frequency of double recombinants in the corresponding three-factor cross. For example, for the adjacent intervals AB and BC, $R_1 + R_2$ is 0.38, the corresponding three-factor cross is

$$a \quad\quad\quad b^+ \quad\quad\quad c$$
$$a^+ \quad\quad\quad b \quad\quad\quad c^+$$

and the observed frequency of double recombinants is .020. Boot-strap your way to a mapping function via steps (a) and (b) below.

(a) From the data, construct a graph of S versus $R_1 + R_2$.

(b) Using this graph, construct a graphic mapping function for this creature.

$R_1 + R_2$	R_{doubles}	$R_1 + R_2$	R_{doubles}
0.05	0.000047	0.69	0.104
0.10	0.00037	0.74	0.127
0.20	0.0029	0.79	0.150
0.38	0.020	0.87	0.186
0.47	0.036	0.93	0.212
0.55	0.057	0.96	0.230
0.62	0.079	0.99	0.245

1-10 Compare the graphs of your mapping functions in Problems 8 and 9. They should be very similar since the data for both problems were concocted using the same algebraic function. If you feel inspired, see if you can create an $R = f(x)$ that will fit your graphs.

NOTES

1. A. H. Sturtevant (1913) *JEZ* **14**:43.
2. L. C. Norkin (1970) *JMB* **51**:633.
3. A. H. Sturtevant and G. W. Beadle (1962) "*An Introduction to Genetics*," New York: Dover (originally published in 1939).
4. J. B. S. Haldane (1919) *JG* **8**:299.
5. K. Mather (1951) "*The Measurement of Linkage in Heredity*," London: Methuen.
6. P. Amati and M. Meselson (1965) *G* **51**:369.

2

Random Particle Analysis

For phages, an assumption of random, sequential, pairwise "matings"
leads to adequate mapping functions. For phage λ, the exchanges in any
given mating may follow Haldane's Function, i.e., they may occur without
interference. For T4, radical departure from Haldane's Function is
demanded for two reasons: the T4 linkage map is circular, and sequential
interactions ("matings") involve only short segments of T4 chromosomes.

THE PHAGE CROSS

The attraction of phages to students of recombination is the relative simplicity of their chemical structure. These viruses that grow in bacteria are
so simple that one can expect them to give answers to questions posed in
chemical terms! Such hopes are now beginning to be realized. In fact, it is
this partial fruition that is the impetus for this book. Like the original
workers in phage genetics, we shall borrow concepts and vocabulary
from eukaryotic genetics. This approach assumes of course that recombinational mechanisms in phage and fungi are sufficiently similar that these
otherwise separate fields of study can profit from an interchange of points
of view. As it turns out, this assumption is justified, even though the extension of concepts has on occasion contributed to confusion.

As in eukaryotic genetics, the cross is the basic operation of the phage
geneticist. It takes a variety of forms, all sufficiently analogous to the concept
of cross in fungal genetics to justify the extension of that concept. A phage
cross is usually conducted as follows. A large number of particles, say 10^9,
of each of two (sometimes more!) genotypes is adsorbed to a smaller number,
say 10^8, of bacterial host cells. The host cell concentration is high to promote
adsorption of phage particles, and the phage concentration is 5–20 times

higher to ensure that almost all cells get infected with comparable numbers of particles of each of the infecting types. As we shall see, the recombinant frequencies depend on the numbers of particles adsorbed per cell for both trivial reasons and, in some phages (T4 for example), for reasons having to do with the mechanism of recombination. After most of the infecting particles have adsorbed, it is usual to dilute the infected culture so that progeny particles released at cell lysis do not adsorb to bacteria or to bacterial debris. The phage progeny produced from a large number (several thousand or more) of such infected cells is then sampled for genotype frequencies.

The mutations widely used for recombination studies in phages are most often of two sorts: conditional-lethal mutations and plaque-morphology mutations. Conditional-lethal mutations render a phage incapable of plaque formation in a normally benign environment while permitting plaque formation in another. (The environments are "nonpermissive" and "permissive," respectively.) Examples of conditional-lethals are the mutations that create "stop translation" codons (*sus*, *am*, etc.), those that render phage growth temperature sensitive (*ts*), and host-range mutations. Plaque-morphology mutations change the size and/or clarity of the plaque (Figure 2-1). The *rII* mutations of phage T4 remain highly useful partly because they are simultaneously conditional-lethal and plaque-morphology mutations.[1]

As with eukaryotes, the frequency of recombinants for a given pair of markers is a reproducible property of that pair of markers. However, reproducibility in phages can be weaker than in eukaryotes. One of the reasons for this is certainly that the recombinant frequency in phages depends on both the average number of particles adsorbed per cell (multiplicity of infection) and the relative amounts of the two infecting genotypes adsorbed (see below).* Reproducibility of recombinant frequencies is good enough, however, to permit the construction of linkage maps.

In general, all the loci identified in a given phage can be mapped in a single linkage group corresponding to the single DNA molecule (or chromosome, to again borrow from the terminology of eukaryotic genetics). The details of these maps and the structures and behaviors of the chromosomes underlying them differ for the different phages. We shall examine two of them individually here and in later chapters.

The first phages examined showed certain common features, indicating that crosses of phages differed in some important respects from crosses of eukaryotes. (1) When infections were made with three different genotypes, some progeny particles contained markers from each of the three kinds of

---

* Notice that we say "recombin*ant* frequency" rather than "recombin*ation* frequency," for reasons to be explained.

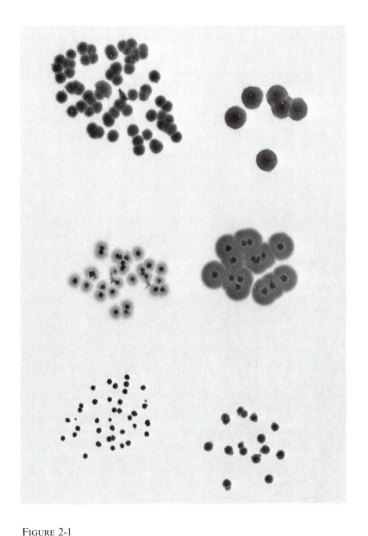

FIGURE 2-1

To assay phage particles, about 10^8 sensitive bacteria are mixed with about 10^2 particles in a total volume of about 10^{-1} ml. About 2 ml of melted nutrient agar is added and the mixture is poured over the surface of a hard nutrient agar in a petri dish. The top layer hardens and the plate is incubated. The 10^8 bacteria give rise to 10^8 colonies so small and tightly packed as to look like a continuous film of cells. This film or "lawn" is interrupted where phage particles have formed plaques by successive rounds of invasion of the cells, growth within them, and subsequent escape by destruction of the cell envelope (lysis). Some phage mutants form plaques that look different from those made by wild-type viruses (*upper left*). The strains of the phage T4 shown above were described and used in the 1950s. (A. H. Doermann [1953] *CSHSQB* **18**:3; photo courtesy of A. H. Doermann.)

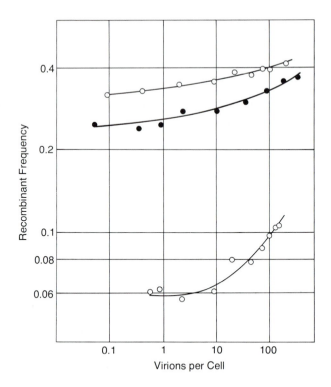

Figure 2-2

Recombinant frequency among intracellular virions (mature phages) for three different crosses as a function of the number of virions per cell. The last point in each curve corresponds to the number of phages released (~ 200) as a result of normal phage-induced lysis. The earlier points were obtained by inducing cell lysis prematurely. (A. H. Doermann [1953] *CSHSQB* **18**:3.)

infecting phages ("triparental recombinants"). (2) When host cells were lysed prematurely, it was observed that the small population of mature phages or virions had lower recombinant frequencies than did the larger populations of phage present in the cell at later times (Figure 2-2). (3) Crosses were characterized by coefficients of coincidence greater than unity. (4) Markers that were unliked by one standard criterion were linked by another. In the three-factor cross $abc \times a^+b^+c^+$, c and c^+ were (about) equally frequent among both the ab^+ and the a^+b recombinants, indicating nonlinkage of C to either A or B. However, in the total progeny, the frequencies of recombinants for A and C or for B and C were unambiguously less than 50 percent. (5) With loosely linked markers, infection by unequal numbers of two genotypes yielded a progeny in which the recombinants were in excess of the minority parent type.

In an apparent effort to unify phage and eukaryotic genetics, the mating theory proposed that a phage cross was analogous to a succession of eukaryotic crosses commencing with the encounter of two genetically distinct populations. As we shall see, this idea was reasonable, economical, and, while shown to be wrong for some phages, remains viable to this day for others.

THE MATING THEORY

The mating theory[2] supposes that the phage chromosomes in a cell are drawn pairwise from a pool to enter a "mating room," the "place" where mating occurs. We need not specify whether the chromosomes that enter the room actually break and exchange parts with each other, whether they specify molecules with exchanges without themselves breaking, whether they increase in number during the process, or what have you. Whatever the mechanisms, the process allows for *some* of the chromosomes that come back out of the room to have enjoyed exchanges, and we say that those chromosomes have had a mating. We suppose that this process follows the specifications of the Model Mark I (Chapter 1).

The mating theory states two rules governing entry into the mating room: chromosomes are selected for entry at random from the mating pool, and, since a chromosome that emerges from the mating room reenters the pool, a single chromosome may be chosen for several matings. These assumptions have several consequences. (1) The fact that chromosome can have more than one mating accounts for the triparental recombinants. (2) The frequency of any recombinant type in the pool rises with the average number of matings; this accounts for the increase in recombinant frequency as a function of time. (3) Chromosomes have different mating experiences; This heterogeneity introduces positive correlations in recombination that appear as coefficients of coincidence greater than unity. (4) Since some chromosomes fail to mate, the recombinant frequency for "unlinked" markers must be less than 0.5. In addition, the mating of two chromosomes that differ by less than two genetic markers will produce no recombinants.

This last feature of the theory calls for a definition of R that is not dependent on the differentiating markers. Instead of noting those markers, we imagine a daub of red stain at each of the two relevant loci (A and B) of one entering chromosome and a daub of green at the same two loci on the other. Any emerging chromosome that has inherited the information at A from the red parent is said to be red at A, etc. Each chromosome gets painted (or repainted) just as it enters the mating room. We define the *descent* of a chromosome according to its color at A. If it emerges from the mating room red at A, then we say it is descended from the chromosome that was daubed red as it entered the room. By keeping an imaginary record of colors, we can define a line of descent for any chromosome; i.e., we can trace its ancestry back to a unique infecting phage particle. Now we define R according to the color of a chromosome at B. If a chromosome from a mating is a different color at B than it is at A, we say it was color-converted at B. The probability of color-conversion in any single mating is R. Since chromosomes get repainted just as they enter a mating, color conversion can occur more than once in the lineage of any phage chromosome. The average number of color conversions per lineage is mR, where m is the average number of matings per lineage. Now we are ready to specify the frequency of one recombinant type, say a^+b^+ in the mating pool of a cross between infecting types a^+b and ab^+. To do so, close your eyes and pick a chromosome from the pool. The chance that this chromosome is a^+ is simply the probability that it is descended from the a^+b parent. This proba- bility is f, the fraction of infecting phage of type a^+b (f is usually, but not always, adjusted to be $\frac{1}{2}$). Now in order for the selected chromosome to be genotype a^+b^+, its lineage must have enjoyed at least one color conversion. If matings last but a moment, then they will be Poisson-distributed among lineages, and the probability that a given lineage has had at least one color conversion is $1 - e^{-mR}$. A second condition for having the genotype a^+b^+ is that the last color conversion in the lineage must occur at the hands of a chromosome carrying the b^+ allele. The probability of this occurrence, if the pool is well mixed, is just $1 - f$, the frequency of the ab^+ parent in the infecting mixture, so that the frequency of the a^+b^+ recombinant is $f(1 - e^{-mR})(1 - f)$. Since the frequency of the a^+b^+ recombinant will equal that of the ab one (Rule 1, Chapter 1), we can write the recombinant frequency, r, in the pool as

$$r = 2f(1 - f)(1 - e^{-mR}) \tag{2-1}$$

which, when $f = \frac{1}{2}$, is

$$r = \tfrac{1}{2}(1 - e^{-mR}) \tag{2-2}$$

Phage mating goes on in *individual* host cells, and the numbers of particles adsorbed per cell is not very large, say from 3–10 of each type in most experiments. Due to the random (approximately Poisson) distribution of infecting particles upon bacterial cells, the realized fractions of infecting phage of each type will vary from cell to cell. The fraction of cells in which

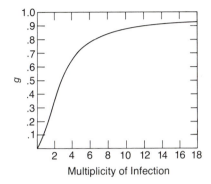

FIGURE 2-3
The finite-input factor g as a function of the total multiplicity of infection (average number of phage particles adsorbed per cell). (E. S. Lennox, C. Levinthal, and F. Smith [1953] G **38**:508.)

f is realized approaches unity only as the multiplicity of infection approaches infinity. The "finite input" of parental phage in actual crosses reduces r below that described by Equation 2-1. The factor g by which recombination is reduced has been expressed (Figure 2-3), and our equation becomes

$$r = 2gf(1 - f)(1 - e^{-mR}) \qquad (2\text{-}3)$$

Equation 2-3 describes the frequency of recombinants in the mating pool of phage chromosomes. Methods for the direct measurement of r are not generally available. Such a method would demand a representative recovery of intracellular, unbroken phage chromosomes and their genotyping, presumably by transfection. Because of the difficulties involved in such an analysis, it is simpler to let nature sample the mating pool. The process of *encapsidation* removes chromosomes from the pool and renders them infectious. Encapsidation prevents the chromosome from engaging in further matings during the cross so that the intracellular population of virions is an accumulation of samples of the mating pool withdrawn over a period of time. If mating and maturation are contemporaneous but independent processes, and both go on at a constant rate—all of which may be true for some phages but is certainly not true for all, as we shall see later— then the frequency of recombinants in the mautre phage population can be found by averaging r as a function of m:

$$\bar{r} = \int_{m_1}^{m_2} \frac{r\,dm}{(m_2 - m_1)} \qquad (2\text{-}4)$$

where m_1 is the average number of matings per lineage when the first chromosomes are encapsidated and m_2 is the average in the pool when the mature population is examined.

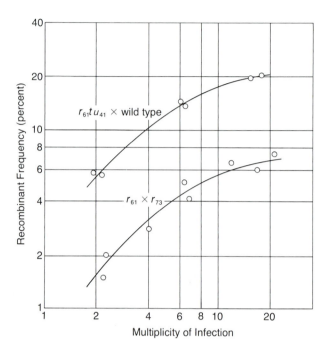

FIGURE 2-4
The dependence of recombinant frequency on multiplicity of
infection in T4. (G. Mosig [1962] *ZV* **93**:280.)

Let us now examine Equation 2-3 to make sure it takes account of the
features of a phage cross described at the beginning of this chapter. The
dependence of r on the relative multiplicity of infection of the two infecting
parents is embodied in the factor $2f(1 - f)$. This feature of the theory is the
easiest to verify and the only one certain to be correct. It is, in fact, a corollary
of the rules of recombination analysis (Rules 1 and 2, Chapter 1) that led
to the concept of a genetic marker. A dependence of r on the total multiplicity
of infection has been described (Figure 2-4). Part of this effect is embodied
in the factor g, which is a recognition of the fact that the average f value
does not describe the situation for all the individual infected cells involved in
the cross. In Chapter 4, when we examine the phage T4 in detail, we shall
note a further dependency of r on multiplicity that is not embodied in
Equation 2-3. The factor $1 - e^{-mR}$ summarizes the remaining features of
a phage cross. The average number of matings per chromosome, m, is a
time-dependent variable and accounts for the increase in r with time. R is
a distance-dependent variable; it accounts for the dependence of r on the
distance between markers.

The relation between R and distance is the interesting knowledge that we hope to squeeze from our interpretations of cross results by means of the mating theory. In particular, are individual phage matings characterized by coefficients of coincidence greater than unity? If so, then phage matings may differ interestingly from the "mating" of eukaryotic chromosomes in meiosis, where S is less than or equal to one (at least for markers that are not very close together.) Or, on the other hand, do R values manifest coefficients like those observed in eukaryotes? If so, we could look even more hopefully to phage genetics for help in elucidating molecular mechanisms of recombination in general. When we come to look at the separate phages λ and T4, we shall see the modest extent to which the mating theory has helped to reveal the nature of the elementary chromosomal interactions. We shall take up λ first because a larger share of its features are reminiscent of eukaryotic recombination as described in Chapter 1. Thus, λ provides a relatively gentle introduction to recombination in bacteriophages.

THE LINKAGE MAP OF LAMBDA
AND THE MATING THEORY

Recombinant frequencies in λ are characterized by coefficients of coincidence (s) greater than unity, a property that is confusingly but justifiably called "negative interference." (We use s for the coefficient of coincidence characterizing a *population* of phages from a cross and reserve S for single "matings", either in phages or in eukaryotes.) Estimates of s based on three-factor crosses are graphed as a function of r in Figure 2-5. In later chapters we shall address the question of rising s values as r approaches zero. Of relevance here is the magnitude of s (~ 3) and its apparent constancy for r greater than about 0.04. It is this sort of negative interference that the mating theory was designed to explain. In qualitative terms, different progeny chromosomes come from lineages with different numbers of matings. These individual matings might be characterized by S of unity or even less than unity. However, the heterogeneity in numbers of matings embodied in the mating theory is supposed to produce the positive correlations in exchanges implied by $s > 1$. For example, if a progeny chromosome from a three-factor cross is observed to be recombinant in one interval, then it must have enjoyed at least one genetically consequential mating. Thus, it is no surprise to find that the particle has a higher than average likelihood of being recombinant in some other interval as well.

The quantitative application of the mating model can be illustrated with data from λ crosses. In principle, the application consists of two steps.

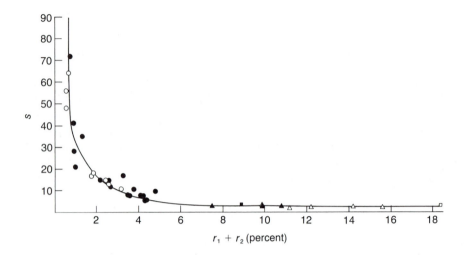

FIGURE 2-5

The relationship of the coefficient of coincidence s to the sum of the component recombinant frequencies in each of 36 different three-factor λ crosses. (P. Amati and M. Meselson [1965] G **51**:369.)

First, a pair of markers that (by some criterion) appears unlinked is identified. In the case of λ, it was observed in the cross $abc \times a^+b^+c^+$ that half of the b^+c class (as well as half of the bc^+ class) was a^+ and the other half a.[3] Thus, A and B appeared unlinked when selection for recombination in the BC interval insured that the chromosomes had enjoyed at least one genetically consequential mating. Therefore, A and B must have an R value close to 0.5. In the overall population, however, the frequency of recombinants, r, between A and B is only ~ 15 percent. Part of the reason for the low r is an apparent limitation on the number of infecting particles that can participate in growth in any single cell. This number has been estimated at five to seven,[4] which sets g at 0.8 (see Figure 2-3). Then, with $r = 0.15$, $g = 0.8$, and $R = 0.5$, Equation 2-3 (which is simpler than and not very different from Equation 2-4) gives $m \simeq 1.0$.

The next step is to transform the r values observed in λ into R values, using the estimate of m. The R values so derived are then tested for the presence of interference by any of the tests of the Haldane Function described in Chapter 1. In fact, the R values found in this way for λ do fit the Haldane Function, suggesting that the negative interference characterizing the r values is a consequence of heterogeneity in the numbers of mating. This outcome permits us to express a mapping function for λ, i.e., a function that

Random Particle Analysis

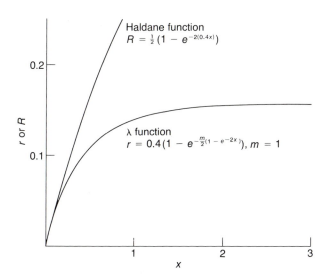

Haldane function
$$R = \tfrac{1}{2}(1 - e^{-2(0.4x)})$$

λ function
$$r = 0.4(1 - e^{-\tfrac{m}{2}(1 - e^{-2x})}),\ m = 1$$

FIGURE 2-6
Graph of Equation 2-5, a mapping function for λ. The recombinant frequency, r, is plotted against the mean number of exchanges, x. Rounds of mating, m, has been set at 1.0. The Haldane Function (Equation 1-2) is graphed for comparison. With both functions, recombinant frequencies vary linearly with exchanges at small values. We have made the graphs coincide in the linear region in order to facilitate their comparison.

relates observed recombinant frequencies with map distances. Combining the mating theory (written for equal multiplicity of the two parents) with the Haldene Function we get

$$r = 0.4(1 - e^{-\tfrac{m}{2}(1 - e^{-2x})}),\ m = 1.0 \tag{2-5}$$

This function is graphed in Figure 2-6 along with the Haldane Function for comparison. To check ourselves and the λ data, we can ask whether our mapping function accounts for the plateau s value of 3 in Figure 2-5. To do this, we solve Equation 2-5 for some convenient value of x and again for $2x$. Then solve for s with Equation 1-6 in the form, $s = (2r_1 - r_2)/2r_1^2$. When I did it for several different x values, I found $s = 3$ in every case. The adequacy of Equation 2-5 as a mapping function for λ has a practical value—it assists us in constructing rectified linkage maps (Figure 2-7). Furthermore, it suggests that λ matings may be rather like those of meiosis

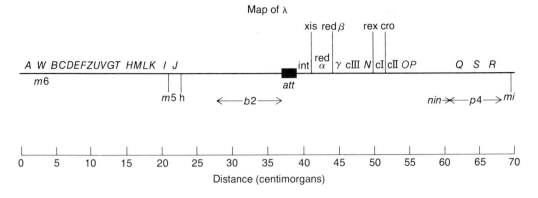

FIGURE 2-7
A rectified linkage map of phage λ. The map shows the position of genes (*above the line*) and several mutations (*below the line*). b_2 and *nin* are deletions; *p4* is a substitution. *att* is the site of action of the Int system (Chapter 4). A centimorgan corresponds to 100 times the *r* values observed for close markers. (A. Campbell [1971] in *The Bacteriophage Lambda*, A. D. Hershey, ed., Cold Spring Harbor Laboratory, p. 13.)

at least when it comes to matters of interference. In Chapters 4 and 10, we shall reveal some problems with this analysis which are better kept hidden for now.

THE LINKAGE MAP OF T4 AND THE MATING THEORY

The map of T4 differs strikingly from those discussed earlier. It has no ends; it is circular instead of linear. What do we mean by a circular map?

> Our "definition" will consist of bounding the conditions for which circular maps are suitable; that is, under what conditions must a conventional map yield to a circular map, and under what conditions is a circular map inadequate?... Consider the (genes) *ABC* shown to be linked in that order. Upon discovery of a fourth (gene), *D*, which also maps between *A* and *C*, the adequacy of the conventional representation becomes testable. That representation is adequate if either of the linkage orders *ADB* or *BDC* can be demonstrated A conventional map becomes inadequate only upon the demonstration of the order *BCD* or *DAB*. A circular map, however, is suitable for this situation and remains so as long as each additional (gene). *N*, which shows linkage *ANC* can be mapped in one of the following positions: *ANB*, *BNC*, *CND*, or *DNA*.[5]

The circularity of a map imposes alterations in our thinking about recombination. First, for a circular map, the recombinant frequency depends not on a single linkage distance between markers but on two. For example, in the cross $ab \times a^+b^+$, recombinant production requires that the linkage between *A* and *B* along both arcs of the circle be broken. A mapping function for T4 must take that into account. Second, the notion of interference loses its usual meaning. In eukaryotes and in λ, as examples of creatures with linear maps, we defined interference (positive or negative) as a departure from a random (Poisson) distribution in the numbers of exchanges within any marked interval among a set of homologous chromosomes from a genetic cross. This definition is somewhat misleading when applied to circular linkage. It is evident from our first point that any recombinant chromosome from a cross yielding a circular map must have enjoyed an *even* number of exchanges. One way to think about this requirement is to recall that recombination requires that each of the two arcs connecting the recombining markers enjoys an *odd* number of exchanges and that the sum of two odd numbers is even. With this warning in mind, we shall talk about "interference" (negative) in T4 crosses, but I shall always put the word and associated terms in quotes to remind us of its somewhat illegitimate use. Its approximate applicability is assured if we apply it only to intervals that are short compared to the total length of the T4 map. With short intervals, the topology of the entire map is of little relevance to any correlations in exchanges within the marked intervals (Figure 2-8). In T4 the "coefficient of coincidence" ("s") calculated on a cross progeny is always greater than unity ("negative interference"), and it rises at the smallest values of *r*. As for λ, in this chapter we shall deal with the modest, more or

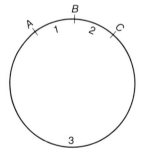

FIGURE 2-8
The commonly employed definition of a "coefficient of coincidence" for circular linkage. The linked genes *ABC* define three arcs on a circular map. In general, two of the arcs (1 and 2 in the diagram) will be shorter than the other. It is conventional (though sometimes confusing) to define a "coefficient of coincidence" as though *ABC* were linked in that order on a linear map. This gives $"s" = \dfrac{r_{AB} + r_{BC} - r_{AC}}{2r_{AB}r_{BC}}$. "Negative interference" is signaled by a "coefficient" greater than one. When arc 3 is far bigger than 1 and 2, the definition for the "coefficient" is insensitive to the map circularity.

less distance-independent, "negative interference" to which the mating theory is addressed.

The obvious explanation for the circularity of a phage map is to suppose that, at least during matings, the chromosome is circular, and such is the case for some phages, e.g., ϕX174. For T4, however, there is no good evidence for chromosome circularity as a regular feature of *any* stage in the life cycle (see Chapter 4). What, then, is the basis for the circularity of the map? Within a clone of T4 particles each chromosome is linear, but each has its genes in a different order! From one particle to another the gene orders are circularly permuted (Figure 2-9). In Chapter 4 we shall discuss the mechanism whereby a single T4 particle can give rise to a clone in which the particles carry chromosomes with different gene sequences. At this juncture we shall simply use the fact in our efforts to construct a "circular" mapping function for T4.

A MAPPING FUNCTION FOR T4

Consider genes A and B on a chromosome of unit length. If all circular permutations of the gene sequence are equally probable, then in a fraction $1 - d$ of the particles the distance separating genes A and B will be d, while in a fraction d of the particles the distance separating genes A and B will be $1 - d$. We shall generate our mapping function within the conceptual framework of the mating theory by supposing that the circular map describes the weighted average linkages in those two subpopulations.[6]

The r values for T4 genes that are one-fourth or even more of the way around the map from each other are detectably less than those for markers halfway around. The number of exchanges per mating must therefore be small, which presents us with an algebraic option. When x is small, the Haldane Function (Equation 1-2) becomes $R = x$, and the basic mating theory function (Equation 2-1) becomes

$$r = 2f(1 - f)(1 - e^{-mx}) \tag{2-6}$$

Since m and x occur as a product in this equation, there is no way to evaluate them separately; if we imagine a vanishingly small number of exchanges per mating, we have only to invoke an immense number of matings (most of them without any exchanges) to compensate. Keeping count of matings in which no exchanges occur seems so foolish that we are inspired to abandon the Haldane Function for R and assume instead that every mating involves exactly one exchange. We may think about this maneuver in either of two ways. Either we redefine a mating such that we count only those with at least one exchange and assume that those with more than one exchange are negligible, or we may suppose that the T4 chromosome is genetically so

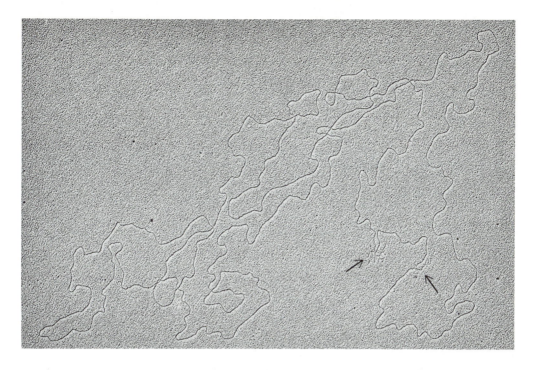

FIGURE 2-9
Electron micrograph of a DNA duplex resulting from the melting and subsequent renaturation of the chromosomes of a clone of T2 virions. Whereas the individual chromosomes of the virions are linear, the annealing of complementary chains from different particles yields a circular structure. This result visually demonstrates the circularly permuted sequences of genes among the virions of a T2 clone. The short single-chain regions indicated by arrows are terminal redundancies (Chapter 4). Phage T2 is closely related to T4. (C. A. Thomas [1967] in *The Neurosciences*, G. C. Quarton, T. Melnechuk, and F. O. Schmitt, eds., Rockefeller University Press, New York; photo courtesy of C. A. Thomas, Jr.)

long as to guarantee an exchange in each mating and that complete interference ($S = 0$) prevents additional exchanges. The wisdom of our inspiration will be measured by the success with which our function transforms r values to (additive) distances.

The assumption that each mating has exactly one exchange leads to $R = d$ for the chromosomes in which A and B are separated by d, and $R = 1 - d$ in the others. Then we can write our T4 mapping function as

$$r = 2gf(1 - f)[(1 - d)(1 - e^{-md}) + d(1 - e^{-m(1-d)})] \qquad (2\text{-}7)$$

Averaging Equation 2-7 to account for spread in maturation times gives

$$\bar{r} = 2gf(1 - f)\left\{(1 - d)\left[1 + \frac{e^{-m_2 d} - e^{-m_1 d}}{(m_2 - m_1)d}\right]\right.$$
$$\left. + d\left[1 + \frac{e^{-m_2(1-d)} - e^{-m_1(1-d)}}{(m_2 - m_1)(1 - d)}\right]\right\} \qquad (2\text{-}8)$$

In Problem 2-10 we shall evaluate m_1 and m_2 as 2.5 and 9.5, respectively, and Equation 2-8 can be tested against the data.

In Figure 2-10, d values obtained by application of Equation 2-8 to observed $\bar{r}$ values are tested for additivity. The fit looks good. Especially encouraging is the 1.06 obtained by summing d values clear around the circle (a value of 1.0 would have indicated a perfect fit of the data to Equation 2-8). We shall defer the presentation of a rectified T4 map until Chapter 6.

Equation 2-8, which fits the data, assumes that there is exactly one exchange in each mating. The apparent adequacy of this assumption casts suspicion upon the very concept of mating in T4. If exchanges occur one in each mating, then perhaps we should dispense altogether with the notion of mating. Matings were originally invoked to account for negative interference by providing opportunities for temporally clustered exchanges. A direct

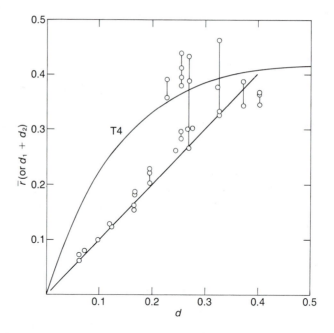

FIGURE 2-10
A mapping function for T4 (*curve*) and a test of its applicability. The curve is a graph of Equation 2-8, with $g = 0.9$, $f = 0.5$, $m_1 = 2.5$, and $m_2 = 9.5$. The points were obtained by using the curve to convert observed $\bar{r}$ values to d values and then testing those d values in pairwise combinations for additivity. Sets of points that share a common estimate of d are connected by vertical lines. The straight line of unit slope is the expectation for perfect additivity. (F. W. Stah, and C. M. Steinberg [1964] *G* **50**:531.)

test of the mating theory has supported the view that T4 chromosomes experience individual exchanges without involving whole chromosomes in a mating.

DIRECT TEST OF THE MATING THEORY FOR T4

The notion basic to the mating theory is that of pairwise entry into a mating room. This notion was challenged by a pair of crosses.[7] The first cross was biparental; i.e., infection was with two genotypes of phage. They carried markers as follows:

$$\underline{a_1 a_2^+} \qquad \underline{b_1 b_2^+} \qquad \text{Parent 1}$$
$$\times$$
$$\underline{a_1^+ a_2} \qquad \underline{b_1^+ b_2} \qquad \text{Parent 2}$$

(Since the circularity of the T4 map has no bearing on this experiment, I use the more easily drawn linear map for illustration.) Both a_1 and a_2 are mutations in one gene; b_1 and b_2 are mutations in a different gene, B, located far from A. The frequencies of $a_1^+ a_2^+$ and $b_1^+ b_2^+$ were measured. Then the cross was varied to include a third parent:

$$\underline{a_1 a_2^+} \qquad \underline{b_1 b_2^+} \qquad \text{Parent 1}$$
$$\underline{a_1^+ a_2} \qquad \underline{b_1^+ b_2} \qquad \text{Parent 2}$$
$$\underline{a_{del}} \qquad \underline{b_1 b_2} \qquad \text{Parent 3}$$

The third parent is deleted for the entire A gene and is double mutant in the B gene. The third parent outnumbered the other two. According to the mating theory, the presence of the third parent will reduce the wild-type recombinants in both the A and the B genes for two reasons: dilution and decoy. Dilution means that the third parent adds to the yield of mutant progeny but cannot contribute to the formation of wild type. Decoy means that parents 1 and 2, each of which selects mating partners at random, will choose parent 3 most of the time because of its high frequency in the pool. Of course the matings with parent three cannot result in wild-type progeny. In the mating theory the two effects are embodied in the factors f and $(1 - f)$, respectively. The death of the mating concept for T4 was established by the differential response of the A and B genes to the presence of the third parent. The B gene, for which the third parent was double mutant, showed both the dilution and the decoy effects. The A gene, however, for which the third parent was deleted, showed only the dilution effect. If entire chromosomes entered the mating room, as specified by the mating theory, then both genes should have shown both effects. Instead, with respect to the *A* gene, parents 1 and 2 act as though the third parent is not even in the pool, and we must conclude that the interactions leading to recombinant formation in T4 do not involve entire chromosomes but instead are confined to short stretches. In Chapter 6 we shall consider the length of those stretches.

Although the mating notion was laid to rest for T4, our mapping function (Equation 2-8), which is a descendent of the mating theory, is not challenged by that result. On the contrary, Equation 2-8 and the experiment described above are harmonious in their conclusion that exchanges occur independently and not in temporal clusters ("matings"). (In Chapter 6, where we shall examine T4 crosses with very close markers, we shall realize that the event we call *an* exchange here may sometimes in fact be a cluster of closely neighboring exchanges.)

The mating theory, which we now consider ruled out for T4, was originally introduced to account for "negative interference" in that very phage. How, then, are we to account for this phenomenon? Judging from the success of our mapping function, "negative interference" in T4 can be accounted for by three factors. First, the circularity of the map contributes "negative interference." Recombination of *A* and *B* implies exchange not only in the short arc connecting *A* and *B* but also in the long arc. Sometimes, of course, this exchange will recombine *B* and *C*, resulting in "negative interference." Therefore, to the extent that "negative interference" is accounted for by this consideration, we may say that it is an artifact of employing an inappropriate definition for the "coefficient of coincidence" ("*s*"). Two other factors, embodied in the equations for the mating theory (Equation 2-8), account for the remainder of the "negative interference." Finite input introduces cell-to-cell and hence lineage-to-lineage heterogeneity in numbers of genetically consequential exchanges, and the variation in maturation time of chromosomes introduces further heterogeneity.

PROBLEMS

2-1 (a) If 1×10^8 bacteria are infected with 3×10^8 particles of phage type *a*, what fraction of the cells receive no particles?

(b) If 3×10^8 of type a^+ are added along with the *a* particles, what fraction of the cells receive no particles of either type?

(c) What fraction of the cells in (b) are infected by both types?

(d) Answer (a), (b), and (c) assuming that the number of *a* and of a^+ particles is each 1×10^9.

2-2 What is the value of the finite-input correction (g) when the total multiplicity of infection is six? When the total is 20?

2-3 When the average number of matings in the pool is three, what fraction of the lineages in the pool have had no matings?

2-4 Assuming equal finite input, calculate the recombinant frequency in the pool when the average number of matings is three for a pair of markers for which color conversion occurs with a probability of 0.2 per mating.

2-5 Jan intended to do a phage cross with equal numbers of two parents, but he goofed. By mistake he added five particles per cell of one type and 15 per cell of the other. The frequency of recombinants among the progeny was 0.082. What recombinant frequency would Jan have seen if he had succeeded in adding 10 particles of each type per cell?

2-6 Gus's phage stocks were rather low titer, so he was able to add only two particles per cell of each type (giving him a total multiplicity of infection of four). He observed a recombinant frequency of 0.2 among the progeny virions. What frequency would he have observed if he had added his usual seven particles of each type per cell? (Assume that the only effect of multiplicity of infection is that of "finite input.")

2-7 If a cross is conducted at a total multiplicity of infection of 15, what is the largest recombinant frequency that can be obtained? That is, what value of r would be observed if the product mR were large?

2-8 A set of crosses was performed with very high equal multiplicities of two parents. The following (concocted) recombinant frequencies were observed:

Cross	r	Cross	r
$A \times B$	0.20	$C \times D$	0.26
$A \times C$	0.28	$C \times E$	0.31
$A \times D$	0.13	$C \times F$	0.31
$A \times E$	0.28	$C \times G$	0.31
$A \times F$	0.30		
$A \times G$	0.29	$D \times E$	0.29
		$D \times F$	0.30
$B \times C$	0.30	$D \times G$	0.29
$B \times D$	0.24		
$B \times E$	0.22	$E \times F$	0.22
$B \times F$	0.28	$E \times G$	0.08
$B \times G$	0.24		
		$F \times G$	0.19

(a) What is the order of the genes A–G?

(b) To what maximum values does the recombinant frequency rise as distances become very large?

(c) Calculate m (assume that $m_1 = m_2$, i.e., that encapsidation occurs after mating is completed.)

(d) Construct a rectified map for this make-believe phage with the aid of Equations 2-2 and 1-2.

2-9 What fraction of the circular map of T4 separates two markers along the short arc if they give 19 percent recombinants in a standard cross? (Answer with the aid of Figure 2-10.)

2-10 In the mating theory for T4 (Equation 2-7), $d = (1 - d) = \frac{1}{2}$ for markers directly opposite on the circular map.

(a) Write Equation 2-7 for the special case of $d = \frac{1}{2}$ and $f = \frac{1}{2}$.

(b) In T4 crosses performed at a multiplicity of infection of 14 $(g = 0.9)$, markers directly opposite on the circular map gave $r = 0.32$ among the first virions to appear in the infected cells. In the equation from (a) solve for m and call the value obtained m_1.

(c) When d is small, Equation 2-7 can be written as

$$r \cong 2gf(1 - f)[md + d(1 - e^{-m})]$$

For m values of the magnitude encountered in T4, this can be simplified to

$$r \cong 2gf(1 - f)d(m + 1)$$

Note that r increases linearly with m for close markers, so that $\bar{r}$, the recombinant frequency in the virion population at normal lysis time, is

$$\bar{r} = 2gf(1 - f)d(\bar{m} + 1)$$

For close markers in T4, $\bar{r}$ is just about twice the r value observed among the first virions (see Figure 2-2). Calculate $\bar{m}$.

(d) By definition, $\bar{m} = (m_2 + m_1)/2$. Calculate m_2.

NOTES

1. S. Benzer (1955) *PNAS* **41**:344.
2. N. Visconti and M. Delbruck (1953) *G* **38**:5; C. Steinberg and F. Stahl (1958) *CSHSQB* **23**:42; A. D. Hershey (1958) *CSHSQB* **23**:19.
3. A. D. Kaiser (1955) *V* **1**:424.
4. E. L. Wollman and F. Jacob (1954) *AIP* **87**:674.
5. F. W. Stahl (1967) *JCP* **70** (Suppl. 1): 1.
6. F. W. Stahl and C. M. Steinberg (1964) *G* **50**:531.
7. J. D. Drake (1967) *PNAS* **58**:962.

3

Meiosis and Tetrad Analysis

Each chromosome has essentially doubled prior to the pairing of homologs in meiosis. The fact that exchange occurs in the four-chromatid stage raises questions concerning which pairs are allowed to exchange and the influence that exchanges might have on each other. Exchanges do interfere with each other along the length of the chromosomes, but the combination of strands involved in one exchange has little influence on the combination involved in neighboring exchanges. Exchanges between sister chromatids probably occur in some eukaryotes, but seem not to in others. The incompleteness of premeiotic DNA replication observed in some organisms may relate to the occurrence and properties of exchanges between sister chromatids.

OUTLINE OF MEIOSIS

In eukaryotes reduction in chromosome number and genetic recombination both occur during meiosis. Thus, a study of chromosomes in meiosis will provide substance to our Model Mark I (Chapter 1) and will lead to its further specification. We shall make no effort to describe even half of what is known or speculated about meiosis; that is done in many other places. Instead, we shall focus on that which seems to have the most immediate bearing on the mechanisms of recombination. The description of meiosis below is a composite one, based mostly on eukaryotes other than fungi.

Early in meiosis, before the chromosomes condense (Figure 3-1), most of the DNA in the nucleus is replicated.[1] In some organisms this replication occurs prior to the fusion of the haploid nuclei that originates the diploid meiocyte.[2] That observation has a powerful bearing on the admissibility of some of the basic recombination schemes proposed—the break-join, copy-choice, and break-copy models (see Figure 1-3). It tells us that in those organisms in which replication occurs prior to fusion, the chromosomes

First Meiotic Division

1. The chromosomes first become visible to ordinary microscopic examination in the leptotene stage. Each chromosome typically appears to be composed of but a single chromatid, though DNA duplication has occurred prior to this stage.

2. Homologous chromosomes undergo pairing (synapsis) in the zygotene stage, producing chromosomal structures called bivalents.

3. In many organisms (but not all of them, for example, the salamanders) each chromosome is visibly composed of two chromatids at the pachytene stage, so that each bivalent can now be seen to consist of four strands of genic material.

4. The chiasmata, which look as if they arose by exchange of homologous parts between synapsed chromatids, are clearly visible in the diplotene stage.

5. Following the orientation of the four-strand bivalents (synapsed pairs of homologous, duplicated chromosomes) on the equatorial plane of the spindle apparatus in metaphase I, the homologous centromeres move toward opposite poles of the spindle in anaphase I, separating the four strands of genic material two-from-two and thus producing two sets of half-bivalents.

6. Reformation of the nuclear membrane is followed by cell division in telophase I. In interkinesis the half-bivalents become elongated; chromosome duplication does not occur during this stage.

Second Meiotic Division

7. The second meiotic division is initiated by shortening and thickening of the half-bivalents in prophase II. This is followed by their migration to the equatorial plane of the spindle apparatus in metaphase II. The two chromatids making up each half-bivalent are separated one-from-one in anaphase II by the movement of sister centromeres toward opposite poles of the spindle.

8. Telophase II results in the formation of four haploid cells, the final products of meiosis.

FIGURE 3-1

A diagram of the behavior of chromosomes during meiosis. (Adapted from *The Mechanics of Inheritance*, 2nd ed., F. W. Stahl. Copyright © 1969 by Prentice-Hall, Englewood Cliffs, N.J.)

that interact to produce recombinants are already replicated over most of their length before they become involved in the exchange process. (In other creatures, it may be that replication and recombination are contemporaneous processes.[3] Perhaps a wider variety of schemes remains admissable in those cases.) Later in meiosis, still in the prophase (Figure 3-1), the exchanges take place.[4] The DNA that was not replicated during the premeiotic chromosome replication is replicated at about this time (in the zygotene stage), and small bits of DNA previously replicated are copied again (in the pachytene stage).[5] These observations suggest that local DNA synthesis may be involved in the exchange process.

As the chromosomes continue to condense in the diplotene stage, each is seen to be double (composed of two chromatids), and the exchanges are manifested as cross-connections (chiasmata) between homologs.[6] The chiasmata slide to the ends of the bivalents before or while the homologous centromeres are directed to opposite poles of the spindle. This first meiotic division is followed by a second one without detectable intervening DNA synthesis (Figure 3-1). The result of the two divisions is four haploid cells that are variously recombinant for any markers differentiating the two fusing haploid nuclei. In ascomycetes, the fungi favored for genetic study, these products of meiosis are retained in one bundle (an ascus), which can be isolated and separated into its component cells by the investigator. In some ascomycetes each haploid nucleus undergoes a mitosis prior to the partition of nuclei into separate cells (ascospores) in the ascus. These eight-spored fungi have special value for the study of recombination between very close markers, as we shall see in Chapters 7 and 8. In some eight-spored fungi, e.g., *Neurospora crassa* (Figure 1-2), the four spore pairs retain an orientation in the ascus that permits identification of cell-pairs that are sisters of the second meiotic division, or *ordered tetrads*.

The determination of genotypes of cross progeny that are members of the same tetrad, ordered or otherwise, is called tetrad analysis. It is this operation with which we shall be concerned in this chapter. Our goal is to uncover the rules that govern the exchanges between the strands resulting from the premeiotic DNA replication. We hope that realization of this goal will suggest features of the exchange mechanisms themselves. At the very least, it will set rules for models that we may try to build later.

THE MODEL MARK II

The bare outline of meiosis presented above already forces an elaboration of the Model Mark I. The elaboration is sufficiently extensive to justify our calling the elaborated model Mark II. The Model Mark II recognizes that chromosomes can replicate either completely (Mark IIA) or incompletely (Mark IIB) prior to exchange. We diagram the alternative starting conditions as:

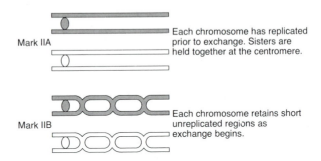

Mark IIA — Each chromosome has replicated prior to exchange. Sisters are held together at the centromere.

Mark IIB — Each chromosome retains short unreplicated regions as exchange begins.

Most of the data with which we shall deal in this chapter lack the power to distinguish between IIA and IIB. Therefore, while recognizing the biochemical support for IIB, we shall work with the diagrammatically simpler IIA. We shall recall IIB whenever the need rises.

Obviously, the variety of ways in which four chromatids can interact during exchange vastly exceeds the possibilities conceivable with only two. Fortunately, a little tetrad analysis allows us to eliminate some of the possibilities. Unfortunately, however, it falls short of providing convincing answers to all of our questions, including some apparently simple ones.

TWO-FACTOR CROSSES

For chromosomes marked at two loci, we diagram the Model Mark II as follows:

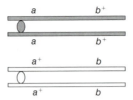

The tetrads resulting from the two meiotic divisions following exchange are observed to be of three kinds (see Figure 3-2). The PD tetrads contain only parental-type spores; the T tetrads contain two recombinant and two parental types; the NPD tetrads contain four recombinant spores.

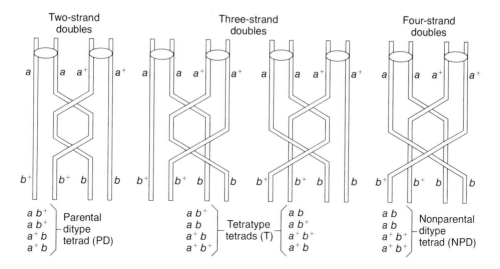

FIGURE 3-2
The manner in which two exchanges can dispose themselves in a bivalent. Assuming that sister-strand crossovers do not occur, four distinguishable arrangements are possible. To illustrate the consequences of double exchange in a three-factor cross, add markers c and c^+ halfway between A and B. (Adapted from *The Mechanics of Inheritance*, 2nd ed., F. W. Stahl. Copyright © 1969 by Prentice-Hall, Englewood Cliffs, N.J.)

The very existence of T tetrads is harmonious with the conclusion, based on measurements of DNA quantity, that exchange follows chromosome replication. In fact, by itself it pretty well rules out the possibility that all exchanges precede replication.[7]*

Highly significant for the elaboration of our model is the finding that genotypes in individual tetrads are almost invariably present in complementary pairs. Each a^+b^+ recombinant is accompanied by an ab recombinant; each a^+b parental type is accompanied by an ab^+ parental type. This finding allows us to reject any version of our model that fails to assure that complementary recombinants are formed simultaneously. (The reality is not entirely invariable because occasionally tetrads are encountered in which complementary types are not in pairs. These violations of reciprocality are

---

* The fact that each chromatid is itself a *double* helix prevents us from stating unequivocally that the existence of T tetrads rules out exchange prior to replication. The duplex structure of each chromatid will return to bedevil us when we look at recombination between very close markers. Problem 3-9 invites you to think about a strong experimental demonstration that exchange involves the breaking and rejoining of chromatids following DNA replication.

important in the study of recombination between very close markers, as we shall see in Chapter 7.) Recombinants arising in complementary pairs are called crossovers. Crossovers arise as a result of an odd number of reciprocal exchanges between marked loci.

We shall return to two-factor crosses later in this chapter, but now let us see how the rules of exchange are further delimited by qualitative aspects of the results of three-factor crosses.

THREE-FACTOR CROSSES

The following three-factor cross

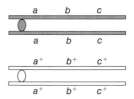

will produce tetrads of the following kinds, among others (we are concerned with the genotypes, not their order in the lists):

Two-strand doubles	Three-strand doubles	Four-strand doubles
a b c	a b c	a b^+c^+
a b^+c	a b^+c^+	a b c^+
a^+b c^+	a^+b c^+	a^+b c
$a^+b^+c^+$	a^+b^+c	a^+b^+c

If we assume that the R values are small enough to be approximately additive, we can say that the above tetrads are two-exchange tetrads. In the two-strand doubles the two exchanges involve only two chromatids. In the three-strand doubles the exchange in interval BC involves only one of the chromatids involved in the AB exchange. In the four-strand doubles the BC exchange involves the two chromatids that were not involved in the AB exchange. The existence of these three different kinds of tetrads tells us that when two exchanges occur in a chromosome interval (AC), they can dispose themselves in any of the three different ways shown in Figure 3-2.

The exchanges diagrammed in Figure 3-2 are those that fit the definition of exchange set down in connection with the Model Mark I. They are those events that result in a change of information source along the chromosome. The Model Mark II suggests that an additional kind of event may occur, one that is physically comparable to an exchange but does not result in a change of information source. These are "exchanges" that might conceivably occur between sister chromatids, i.e., between chromatids that are sister products of the premeiotic replication. Later in this chapter we shall address the question of whether sister-strand exchanges occur. It is a question that is beguilingly subtle and may be important. The subtlety comes from the genetically cryptic nature of such supposed events, since exchanges between genetically identical structures can have no direct effect on the recombination frequency. The importance of the question lies in what the answer could tell us about our Model Mark II. Except for the pairwise way in which the chromatids (strands) are associated with centromeres, Mark IIA shows the four chromatids as equivalent structures. If exchanges are in fact limited to nonsister pairs, or if sister-strand exchange is in any way physically different from nonsister exchanges, then we appear forced to make one of two assumptions. Either the centromeres have a mysterious long-distance effect on exchange, or Mark IIA is wrong and Mark IIB, or something like it in which sisters are acknowledged to have a different relationship than nonsisters, is the preferred model. For now, however, we set the question aside and proceed on the basis of the simplifying (and for many purposes adequate) assumption that sister exchanges do not occur.

In Chapter 1 we noted that exchanges are not Poisson-distributed in most eukaryotes. Instead, they exhibit (positive) interference; i.e., one exchange tends to inhibit others from occurring nearby. In the Model Mark II we must consider two possible origins for this interference.

Chromatid Interference

One possibility is that on any given interval exchanges *are* Poisson-distributed among paired, replicated chromosomes (bivalents). However, the combination of strands involved in one exchange might influence the combination involved in any nearby exchange. Consider the bivalent marked at three loci:

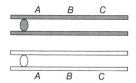

If exchanges are Poisson-distributed among bivalents, then the fraction of tetrads in which recombination occurs in both the AB and BC intervals will be that expected from randomness. However, if neighboring exchanges tend to involve different chromatids, then the frequency of double-recombinant spores will be less than the product $R_{AB}R_{BC}$, and S will be less than unity.

CHIASMA INTERFERENCE

The concept of chiasma interference supposes that one exchange suppresses neighboring exchanges in that bivalent. This interference would be manifest microscopically as a non-Poisson distribution of chiasmata. A straight-forward distinction between chiasma and chromatid interference can be made by tetrad analysis of three-factor crosses. If there is no chromatid interference, then the relative frequencies of two-, three-, and four-strand double-exchange tetrads will be 1:2:1 (see Figure 3-2). This ratio will hold regardless of the presence or absence of chiasma interference. The presence of positive chromatid interference would be signaled by a relative shortage of two-strand doubles; negative chromatid interference would show up as an excess of two-strand doubles. Data from various organisms are generally in accord with the 1:2:1 expectation, with a slight excess of two-strand doubles commonly observed. (We shall return to the two-strand doubles when we deal with recombination between very close markers in Chapter 6.)

There is direct evidence that most interference is accounted for by chiasma interference. (1) For eukaryotes with easily observed chromosomes (which does not include fungi), the microscopically observed distribution of chiasmata among bivalents has been compared with the Poisson distribution. For Poisson's distribution the variance equals the mean; for chiasmata the variance is less than the mean. (2) With enough genetic markers, all exchanges of a bivalent can be enumerated in each tetrad. Again, the distribution in number of exchanges is found to be narrower than Poisson's distribution. Identification of interference as a nonrandom distribution of exchanges among and along bivalents further defines the phenomenon but does not seem to help us reach any view as to its physical or chemical origin.

A feature of two-factor cross data, described in Chapter 1, is that the maximum observed recombination frequency is 0.5. Does that observation jibe with our conclusions about interference? Indeed, it does so nicely. In Chapter 1 I pulled a bit of a fast one; I introduced the concepts of interference and $R_{max} = 0.5$ without acknowledging their possible incompatibility within some conceivable versions of the Model Mark I. Suppose exchange occurred between just two structures, e.g., between homologous chromosomes prior to the premeiotic replication. Now consider the case of two markers far enough apart to assure an exchange between them but with a condition of positive interference so strong that more than one exchange in the interval was a rarity. Such markers would manifest greater than 50 percent recombination! The Model Mark II provides an explanation for the universal failure to observe such R values. An examination of Figure 3-2 reveals that two- and four-strand doubles are equally likely in the absence of chromatid interference. In order for R to exceed 0.5 (for markers in genes A and C), chromatid interference would have to operate so as to make four-strand doubles exceed two-strand doubles. Thus, crossing-over after chromosome replication plus the absence of chromatid interference guarantees $R_{\max} \leqslant 0.5$.

THE HALDANE FUNCTION FOR EXCHANGE IN THE FOUR-STRAND STAGE

These considerations allow us to write a relationship between the recombination frequency and the probability of (nonsister) exchange between two marked loci in a bivalent:

$$R = \tfrac{1}{2}(\text{probability of} \geqslant 1 \text{ exchange}) \tag{3-1}$$

For the special case of a Poisson distribution of exchanges, i.e., no chiasma interference, the parenthetical probability is

$$P_{\geqslant 1} = 1 - e^{-X} \tag{3-2}$$

where X is the mean number of exchanges per bivalent in the marked interval. Combining Equations 3-1 and 3-2 gives

$$R = \tfrac{1}{2}(1 - e^{-X}) \tag{3-3}$$

For exchange in the four-strand stage, the mean number of exchanges per pair of strands (x) is half the mean number of exchanges per bivalent (X) (because there are two pairs of strands per bivalent), allowing us to write

$$R = \tfrac{1}{2}(1 - e^{-2x})$$

which you will recognize as the Haldane Function (Equation 1-2). Thus, our Model Mark II operating without chromatid or chiasma interference leads to the same mapping function as the two-stranded model Mark I operating without interference. General treatments of the four-stranded model that accommodate both chromatid and chiasma interference have been presented.[7]

SISTER-STRAND EXCHANGE

We can now confront the subtle problem of sister-strand exchange. First, we might as well acknowledge that sister-strand exchanges can occur, at least in some situations. They have been shown by autoradiographic methods to occur both in mitosis[8] and in meiosis.[9] Whether or not that evidence, collected in the presence of radiation or chemicals that may have induced the exchanges, is relevant, it does suggest that we alter our strategy and ask the following question. If sister-strand exchange occurs (normally) in meiosis, does it occur on the same basis as nonsister exchange? A negative answer to this question would, by itself, recommend Mark IIB as the model of choice for the present stage of our analysis.

Several lines of evidence bear on the question of sister-strand exchanges. One is based on tetrad analysis of two-factor crosses in fungi. The others come from *Drosophila* and maize.

Given a set of tetrads in which exactly one exchange has occurred in each bivalent between a pair of marked loci, the frequencies of tetrad types will depend on whether or not some of the exchanges are sister-strand exchanges. Although the desired situation cannot be achieved exactly, it might be approximated in an organism that has sufficiently strong chiasma interference. In that case, we can hope to find a pair of loci far enough apart to assure an exchange between them in almost every bivalent, with additional exchanges blocked by interference. Now, suppose that chiasma interference characterizes sister exchanges in the same way as nonsister exchanges, and then consider the consequences of two contrary assumptions: (1) sister exchanges do not occur, or (2) sister exchanges are $\frac{1}{3}$ of all exchanges, where $\frac{1}{3}$ is simply the fraction of chromatid pairs that are sisters. Now, for the well-placed loci described, case 1 would result in all tetrads being tetratype, while case 2 would give no more than $\frac{2}{3}$ tetratype tetrads. In fact, marker pairs are found in fungi that give more than 67 percent tetratype tetrads.[10] This tells us that sister exchanges occur less often than predicted and/or do not manifest chiasma interference to the same extent as nonsister exchanges. A minimal justified statement seems to be the very one we were looking for: sister exchanges do not occur on the same basis as nonsister exchanges.

Bar mutations in *Drosophila* arise by faulty exchange in a certain part of a chromosome. This unequal crossing-over results in a chromatid bearing a duplication that is manifested phenotypically as a dominant mutation influencing the shape of the eye. The involvement of the exchange process in the origin of the mutation is indicated by the high frequency with which newly arising mutants are recombinant for markers bracketing the bar locus. The finding of almost 100 percent recombination among those mutants answers our question about sister-strand exchange.[11] If sister exchanges were as frequent as predicted, i.e., $\frac{1}{3}$ of all exchanges, and were mutagenic, then bar mutants that were not recombinant for the flanking markers should have been found. The paucity of nonrecombinant mutant gametes tells us that either (1) sister-strand exchange occurs at a much lower rate than nonsister exchanges, or (2) sister exchanges never occur unequally, or (3) sister exchange is always accompanied by a nonsister exchange. In any event, we are again led to conclude that sister chromatids relate to each other differently than do nonsisters and that the Model Mark IIB is a useful one for rationalizing this special relationship.

Exchanges between the sister members of a ring chromosome can lead to trouble in anaphase II of meiosis. If the number of sister exchanges is odd, then the two centromeres directed to opposite poles in anaphase II will be connected by a bridge of chromatin. Cytological detection of such bridges in plants[12] argues for a rate of sister-strand exchange at least comparable to the nonsister rate. However, less direct experiments in *Drosophila* failed to find evidence of sister exchange in ring chromosomes.[13]

PROBLEMS

3-1 Haploid nuclei of types ab^+ and a^+b fused to form the diploid ab^+/a^+b. The diploid was cultured, and some of the cells then underwent meiosis. Individual tetrads were examined.

   (a) How many of the cells in each tetrad were type a? How many were type b?

   (b) A tetrad containing exactly one a^+b^+ spore was found. What were the genotypes of the other three spores in that tetrad? What do we call such a tetrad?

   (c) Write the genotypes of the four spores in a parental ditype (PD) and a nonparental ditype (NPD) tetrad, respectively.

3-2 (a) Suppose you had but one marked gene (A) in an organism with unordered tetrads, such as yeast. Convince yourself that examination of tetrads from heterozygous diploids could not determine whether the marked gene was close or distant from its centromere.

   (b) Suppose you had two marked genes (A and B) on separate chromosomes in yeast. What would be the frequency of T tetrads if both genes were remote from their respective centromeres? If one gene were remote but the other gene was very close to its centromere? If both genes were very close to their respective centromeres?

   (c) In (b) we deduced that centromere linkage can be detected in unordered tetrads when we have two unlinked marked genes each close to its respective centromere. Suppose you had such markers and then identified a third marker. What two-factor cross would you conduct to see whether this new marker was close to or far from its centromere?

3-3 Consider the diploid in Problem 1.

    (a) If genes A and B were inseparably linked, what fractions of the tetrads would be PD, T, and NPD, respectively?

    (b) If $R_{AB} = 0.01$, what fractions of the tetrads are PD, T, and NPD, respectively?

    (c) If R_{AB} is 0.40, and 0.70 of the tetrads are T, what fraction of the tetrads are NPD? What is the approximate map distance (x) between genes A and B?

3-4 Consider three close genes linked in the order ABC and the diploid cells $abc/a^+b^+c^+$. Individual meiotic tetrads from these diploids are examined for the rare simultaneous presence of exchange between A and B and between B and C. If there is no chromatid interference, i.e., if the choice of strands involved in one exchange has no influence on the choice involved in another, what fraction of the double-exchange tetrads will be PD for genes A and C, what fraction T for genes A and C, and what fraction NPD for genes A and C?

3-5 Consider marked genes A and B on the same chromosome, and make the conventional assumptions of no sister-strand exchange and no chromatid interference.

    (a) If exactly one exchange between the two markers occurred in each of a number of bivalents, what fractions of the resulting tetrads would be PD, T, NPD?

    (b) If exactly two exchanges occurred, what fractions of the tetrads would be PD, T, NPD?

    (c) If exactly three occurred?

    (d) If an infinite number occurred?

    (e) What fraction would be T if exactly n occurred?

3-6 Suppose n in Problem 3-5e were a Poisson-distributed variable. Call the mean of the distribution X and write an expression for the fraction of T tetrads.

3-7 Using your equation from Problem 3-6, make a graph of T versus X.

3-8 On the graph in Problem 3-7 locate the point relating T and X that you worked out in Problem 3-3c. (Remember that $X = 2x$ since two exchanges in a bivalent make one exchange per chromatid.)

3-9 Chromosomes are labeled with ^{3}H-thymidine prior to the mitosis that precedes the premeiotic DNA replication (W. J. Peacock (1970) G **65**:593). Each chromosome in the diploid cells replicates (the chromosome is drawn with two parallel lines to indicate the duplex nature of the DNA; the heavy lines indicate radioactive polynucleotide chains):

Replication in
^{3}H thymidine

These daughter chromosomes are distributed by the act of mitosis into separate daughter cells. At the premeiotic DNA replication each chromosome again replicates and the daughters (now called sister chromatids) stay connected:

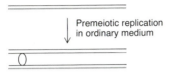

Then homologous chromosomes pair

and exchanges occur.

In cases where no exchange occurs, the chromosomes are seen to be unpaired in squashed cells of meiotic metaphase I. In those chromosomes that are seen to be paired, exactly one chiasma can be seen in each. The chromosomes are sufficiently distinct that each chromatid can be traced along its length and the presence of label along it assessed by autoradiography. Paired chromosomes with a chiasma look like this when exchange has occurred between a labeled and a nonlabeled chromatid:

Each chromosome is composed of two chromatids that are labeled to the same degree at their tips (terminally isolabeled).

(a) If each chiasma is in fact a break-join event, what fraction of the bivalents will show terminal isolabeling?

(b) What coefficients of coincidence would you expect to see in this organism?

(Autoradiographic consequences of sister-strand exchanges were observed in these experiments. By several criteria they were not occurring on the same basis as nonsister exchanges, and they may have been induced to occur by the radioactive label.)

NOTES

1. W. J. Peacock (1970) *G* **65**; 593.
2. J. M. Rossen and M. Westergaard (1966) *CRTCL* **35**:233.
3. R. F. Grell and A. C. Chandley (1965) *PNAS* **53**:1340; R. F. Grell and J. W. Day (1974) in "*Mechanisms in Recombination*," R. F. Grell, ed., New York: Plenum, p. 327.
4. W. J. Peacock (1970) *G* **65**:593.
5. H. Stern and Y. Hotta (1973) *ARG* **7**:37; *G* **78**:227.
6. W. J. Peacock (1970) *G* **65**:593; J. H. Taylor (1965) *JCB* **25**:57.
7. A. Weinstein (1936) *G* **21**:155.
8. J. H. Taylor (1958) *G* **43**:515.
9. W. J. Peacock (1970) *G* **65**:593.
10. R. W. Barratt, D. Newmeyer, D. D. Perkins, and L. Garnjobst, (1954) *AG* **6**:1.
11. A. H. Sturtevant (1928) *G* **13**:401.
12. D. Schwartz (1953) *G* **38**:251; A. Michaelis (1959) *C* **10**:144; J. Miles (1970) PhD Thesis, Indiana University.
13. L. V. Morgan (1933) *G* **18**:250.

4

Phage Replication and Single-Burst Analysis

A study of recombination in phages is impossible without also examining chromosome replication. Phage recombination and replication are contemporaneous, but they are also interrelated in other ways. Furthermore, chromosome encapsidation is dependent on both processes. These relationships undermine basic kinetic assumptions of the mating theory (Chapter 2).

In most cases phage recombination appears to be nonreciprocal. Exceptions are the site-specific Int and the generalized Rec systems acting on λ. The former is evidently a reciprocal break-join system and the latter appears to be so in some circumstances. The Red system of λ, which appears to be nonreciprocal, both enhances and is enhanced by DNA replication and may not be a break-join system.

PHAGE GROWTH

In outline, the replication cycles of λ and T4 are the same. Soon after adsorbtion of a particle to a host cell the chromosome is dissociated from the protein components or capsid of the virion. "Early" genes then direct the synthesis of enzymes that facilitate chromosome replication. As replication proceeds, the expression of "late" genes, whose products are involved in the subsequent process of encapsidation, is exalted by the increase in the number of chromosomes. Genetic recombination goes on contemporaneously with this chromosome replication. Association of chromosomes with newly formed capsids (encapsidation) then begins, with the number of unencapsidated chromosomes being maintained by continued replication. By the time of cell lysis, most of the phage DNA is safely inside the capsids.

Though all phages follow this general scheme, the details vary widely. It is not our intent to record all these differences, which are of interest in themselves. Instead we shall elaborate only those that seem to bear rather directly on genetic recombination in individual phages discussed later in the chapter.

The events in each infected cell are experimentally analogous to an act of meiosis. For a phage geneticist the yield from an individual infected cell is the nearest thing to a tetrad. The mating theory (Chapter 2) asserts homology between a phage mating and meiosis. We can think in terms of matings, but we cannot isolate them or their products for study. Thus, operationally, we have to make do with yields from single cells or *single bursts*. The study of single bursts has been brought to bear on fundamental features of phage recombination.

IS PHAGE RECOMBINATION RECIPROCAL?

We noted that individual tetrads in fungi almost invariably contain complementary genotypes. We deduced from that observation that exchanges in fungi are reciprocal (but also noted that the reciprocality of recombination fails to hold for very close markers). Single bursts have been studied to decide the issue of reciprocality for phages. Two-factor crosses have established reciprocality beyond doubt for fungi. However, two-factor cross data for phages are somewhat less than convincing for two reasons. (1) Even if exchanges are reciprocal, the possibility of unequal replication of the crossovers prior to encapsidation plus the possibility that not all progeny of both crossovers will become encapsidated might diminish their correlated appearance within single bursts. (2) Even if exchanges are nonreciprocal, cell-to-cell heterogeneity (e.g., finite input, Chapter 2) in the opportunities for recombinant formation would introduce correlations in the frequencies of recombinant types within single bursts. However, three-factor crosses can mitigate these uncertainties.[1]

Two parents marked for three loci, say abc and $a^+b^+c^+$, are adsorbed to bacteria as in a standard phage cross (Chapter 2). The culture is then highly diluted and a small constant volume is delivered into each of many tubes. The dilution and volume are adjusted so that only a minor fraction of the tubes receive an infected cell. Some tubes will receive more than one cell according to Poisson's distribution, but the fraction of such tubes among those receiving any cells can be made small. The tubes are incubated until the infected cells have "burst," and the contents of the tubes are then analyzed. In those tubes that have bursts, the frequency of each genotype is determined. Correlations are then calculated for the various recombinant types. As discussed above, these correlations are invariably slightly positive, whatever pair of recombinant types is chosen, because of cell-to-cell heterogeneity. The issue of reciprocality of exchange is confronted by comparing the correlations shown by complementary recombinant types within single bursts and those shown by noncomplementary types. For instance, if ab^+c^+ correlates more strongly with a^+bc than it does with abc^+, then reciprocity of at least some exchanges is indicated. If the former correlation is not stronger than the latter, however, then the method has failed to find evidence for reciprocity, and one might speculate that it does not exist.

Tests for reciprocality of recombination have been made for several phages, not always exactly in the manner described. The result in each case, with the notable exception of the rather special experiments with λ described later in the chapter, has been a failure to demonstrate reciprocity of exchange.

WHEN DOES PHAGE RECOMBINATION OCCUR?

Single bursts have been used to gather evidence on the time in the growth cycle during which recombination occurs. More precisely, the studies deal with questions concerning the temporal relationships between recombination and chromosome replication. Does recombination in each lineage occur exclusively prior to the onset of replication? Is recombination a terminal event occurring just prior to encapsidation so as to preclude subsequent replication? Are recombination and replication not only contemporaneous in the growth cycle but also interdigitated within any given lineage?

Efforts to distinguish the three possibilities have been fashioned after studies on mutant clone size distributions in single bursts. Those studies, using the phage T2, showed about half the mutant clones to be size 1, one-fourth to be size 2–3, one-eighth to be size 4–8, etc. This distribution was like that seen for mutants in a set of bacterial cultures and led to a view of the kinetics of chromosome replication in phages as exponential,[2] or quasi-exponential.[3] Let us assume that a rare exchange (rare because of either small R or small m in Equation 2-3) behaves in some ways like a mutation. If exchanges, like mutations, occur with constant small probability per act of chromosome replication, then the distribution of recombinants should be like that of mutants. (This contention assumes as well that encapsidation and recombination are no differently related to each other than are encapsidation and mutation. We shall see later that this assumption is not always valid.) On the other hand, if recombination in any lineage occurs only prior to replication, and if the recombinants invariably replicate, then recombinants will occur only in large clones. If recombination occurs late, as a last activity prior to encapsidation, then recombinants will be individually distributed among single bursts rather than occurring in larger clones. Each of these models makes a number of assumptions, and the analysis of clone-size distributions has rarely yielded a convincing discrimination among them. The difficulty is one of assessing the origin of a nonrandom distribution. According to the models, a number of recombinants in a single burst is supposed to represent a clone (when r is sufficiently small that *most* bursts contain no recombinants). However, the putative clone could in reality represent recombinants grouped not by filial ties but instead by cell-to-cell variability in the likelihood of recombinant formation. When we examine T4 in detail we shall see what a problem that can be.

RECOMBINATION AND THE CHROMOSOME CYCLE IN LAMBDA

The λ chromosome as it exists in the virion is a linear Watson-Crick duplex of 5×10^4 base pairs. Twelve nucleotides are missing from the 3'OH end of each chain, and the two 5'OH single-chain ends are complementary to each other. Following injection of the chromosome, the complementary ("sticky") ends form a duplex (the *cos* site) circularizing the chromosome (Figure 4-1). The circle becomes covalently closed by the action of the bacterial enzyme polynucleotide ligase. Expression of genes O and P yields proteins that cooperate with preexisting bacterial proteins to replicate the λ circles. Replication is in the so-called theta mode; i.e., two circles are formed in each act. After one or a few rounds of theta replication, the mode of chromosome replication changes to sigma, in which some of the circles become "rolling circles." This transition is aided by the *red* and *gam* genes of λ (see below).[4] Continuous operation of the replication enzymes at the replicating fork of the rolling circle generates a "tail" on the circle.[5] By this time, expression of the "late" genes of λ has provided capsid proteins, and λ chromosomes are cut and packaged from the tails of the rolling circles as they emerge. The encapsidation apparatus ignores any monomeric circles, replicated or not. Apparently, cutting at *cos* and packaging are effectively coordinated only when the DNA to be packaged is part of a DNA structure bigger than the λ chromosome.[6] Recombination, of course, occurs sometime between injection and encapsidation.

If recombination occurs primarily after the onset of rolling-circle replication, then we may have to amend the mating theory analysis of Chapter 2. One study suggests that as few as one to five λ chromosomes per cell make the transition from theta to sigma replication.[7] The cell-to-cell heterogeneity

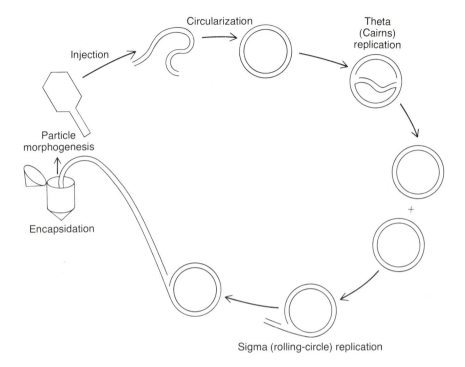

FIGURE 4-1
Replication cycle of the λ chromosome. The linear, duplex chromosome of the λ virion is injected into the host cell. The chromosome circularizes via its complementary single-chain ends. The products of λ genes *O* and *P*, in cooperation with *E. coli* DNA replication enzymes, catalyze replication of the circle, producing a small population of circles. Replication then changes to the rolling-circle mode, the product of which is a serial array of linear λ chromosomes. From this concatemer, chromosomes are cut and packaged into virions to give complementary single chain-ends like those on the original infecting chromosome.

in the ratios of genotypes is greater after the transition than one would calculate from the multiplicity of infection. Instead of $g = 0.8$, as we used in Equation 2-5, a lower value of g might be more appropriate. The possibility that low values for g might sometimes be appropriate is suggested as well by the data in Problem 4-5. Some of the single-burst data shown there reveal highly variable ratios of allele frequencies from burst to burst. In Problems 4-6 and 10-3 we reexamine the mapping function under the supposition that g is small.

Single bursts in λ show little correlation between complementary recombinant types,[8] suggesting that exchange is usually nonreciprocal. However, the issue of reciprocality in λ is not so easily dismissed. In λ three more or less independent enzyme systems—Int, Rec, and Red—effect recombination,[9] so it behooves us to address the reciprocality question separately for each of them.

The Int system acts only upon *att* (Figure 2-7) to recombine circular λ DNA into the host *E. coli* chromosome during lysogenization. During lytic phage growth, the condition of standard crosses (and single bursts), Int can effect exchange between λ chromosomes, again exclusively at *att*. Int appears to be an atypical recombination system with regard to both its site specificity and its mechanism of action. However, for our present purposes it provides a valuable basis for comparison, since some of its features can be convincingly argued from a variety of facts. In Chapter 9 we shall again examine Int; its place in that chapter is warranted by the success of studies of its in vitro activity.

The Rec system is the recombination system of *E. coli*, the host cell of λ. This system is the first prokaryotic one for which defective mutants were found,[10] and its study continues to play a leading role in efforts to elucidate the mechanisms of recombination. One of its components is a nuclease[11] specified by the joint action of the $recB^+$ and $recC^+$ genes, the "$recBC^+$ nuclease,"[12] which interferes with the transition of the λ chromosome cycle into the rolling-circle phase.[13] The *gam* gene of λ informs a protein, the "gamma protein," which inactivates at least partially the $recBC^+$ nuclease, thereby facilitating the transition. The relevant activity of the other clearly identified genetic component of the Rec system, $recA^+$, is unknown (but see Chapter 9).

The Red system, the product of two genes, $red\alpha^+$ and $red\beta^+$, is the major recombination system of λ. When λ is grown in *E. coli* that are severly deficient in recombination, the Red system assures an almost normal frequency of recombinants among progeny particles. The Red system also plays a part in rolling-circle formation in ways we shall speculate upon in Chapter 10.[14] Because of this activity, and because the encapsidation apparatus of λ requires a concatemeric substrate, λ chromosome packaging is not independent of the Red system, and we may speculate that it is not independent of Red-mediated exchange either. The roles of Red and *gam* in the production of concatemeric DNA lead to an interesting relationship of these two systems with the Rec system. In *red gam* infections the transition to rolling circles is not made. Mature phage are produced, however, because the Int and Rec systems promote exchange between the monomeric circles to produce dimers. These dimers are circular for the case of Int (see below), and probably also for Rec, judging from its reciprocality (see below).

If Int and Rec activities are removed by mutation, we may expect a *red gam* infection to produce intracellular DNA but no mature progeny. Indeed, such is the case.[15] In fact, the contribution made by Int is so small that the rather small plaque made by a *red gam* phage is invisible on a *recA* host.[16] On a *recA recB* host plaques appear because mutational removal of the $recBC^+$ nuclease partially restores the ability to form rolling circles.

The study of recombination in λ is made somewhat difficult by these interactions between the recombination systems and by the special relationships of each of the systems to chromosome replication and encapsidation. While bemoaning this situation, we should be alert to the possibility that it may turn out to be a fairly general feature of recombination in most organisms. Now let us examine reciprocality in the three systems and try to deduce the stage in the chromosome cycle at which each acts to effect recombination.

Recovery of Int-promoted dimeric circles from infected cells demonstrates two properties of Int: it acts on circular λ and it does so reciprocally.[17] The best substrate for Int is the nonreplicating circle.[18] This makes fine sense in view of the chaos that would result were Int to act upon any other chromosome form during lysogenization. Not only does Int act upon nonreplicating circles, but it does so whether or not the circles have been allowed any round of replication in that host. Single bursts of $red^+ int^+$ phage in *recA* cells were conducted with phages so marked that Red-mediated recombination in one interval could be distinguished from (predominantly) Int-mediated recombination in another. A stronger correlation was found for complementary Int-mediated recombinants than for Red-mediated ones, supporting the notion of reciprocality in Int.[19]

Fully proper tests for reciprocality in single bursts have been conducted for the Rec system (*red int* lambda in rec^+ cells) and the Red system (red^+ *int* lambda in *recA* cells).[20] Rec was convincingly reciprocal, but Red showed no evidence of reciprocality whatever. It appears as though these two generalized systems may be fundamentally different from each other. However, that conclusion should not be reached on the basis of this evidence alone. Red and, quite possibly, Rec play roles in λ replication that are interdependent (to an ill-known degree) with their respective roles in recombination. Thus, the regimes of replication and encapsidation that await freshly arising recombinants differ in Rec$^-$ and Red$^-$ crosses. These differing regimes could favor encapsidation of similar numbers of reciprocally arising complementary recombinants in one case and make it unlikely in the other.

Suppose, as seems to be the case, that Red recombination occurs primarily after the onset of encapsidation.[21] Suppose furthermore that Red is perfectly reciprocal, giving rise to circular dimers. Now suppose that these dimers are almost always acted upon by the encapsidation system before they replicate. Since only one λ chromosome can be encapsidated from a given dimer circle,[22] one of the two complementary recombinants will be left behind. Two contrasting fates can befall the overlooked recombinant. It may never encapsidate (thus obscuring reciprocality) or it may enjoy a replicational bonanza as a rolling circle and leave a whole parcel of maturable offspring (again obscuring reciprocality but in the opposite sense).

Now suppose Rec is also perfectly reciprocal but acts rather early, before the onset of encapsidation. The resulting dimer will replicate several times before encapsidation starts. Encapsidation of one λ from each member of this clone will probably result in several of each complementary type being represented among progeny phage; evidence of the reciprocality will thus be preserved. The reduction in rolling-circle replication in *red* infections will reduce the chance of replicational bonanzas, thereby further preserving evidence of reciprocality of the recombination event. In Chapter 10 I will offer the proposal that Red is not in fact a reciprocal system—but let the reader beware that the proposal is shaky.

The clone-size distributions of recombinants in single-burst experiments gives some information on the time of action of the recombination systems. As described in the early part of this chapter, the chromosome cycle of λ is itself dependent upon recombination functions. Efforts to learn about recombination from clone-size distributions are thus confounded. However, where recombination systems can be compared in a common background, some suggestive information can be obtained. In single-burst experiments comparing Int and Red in an *int*[+] *red*[+] *recA* infection, the clones for Int recombinants were larger than those for Red.[23] This is compatible with the view here that Int acts on early (circular) forms, and suggests that Red acts later. The suggestion that Red acts on late forms is supported by other studies described later in this section.

If replication is blocked from the beginning of an infection, then the injected chromosomes are held in a circular state.[24] Despite this block, capsid proteins are produced and those chromosomes that have become dimeric by exchange can be encapsidated.[25] One can examine the exchange distribution along the λ chromosome in the following kind of replication-blocked cross. One parent is density labeled and the other not. The two parents differ by markers in the terminal genes *A* and *R*. If single exchanges outnumber multiples, then the density distribution of genetic recombinants will reveal the exchange distribution along the chromosome (Figure 4-2). Furthermore, since dimer formation by exchange is prerequisite to encapsidation, we may expect the total yield of phage to reflect the total rate of recombination. This salubrious situation can be used to study the action of the several recombination systems on the early, circular chromosome of λ. For heuristic reasons it is best we start with an infection made by *red int* phage in which the action of Rec is revealed.

During λ infection the Rec system is partially blocked by *gam* inhibition of the *recBC* nuclease. (Residual recombinational activity is probably due

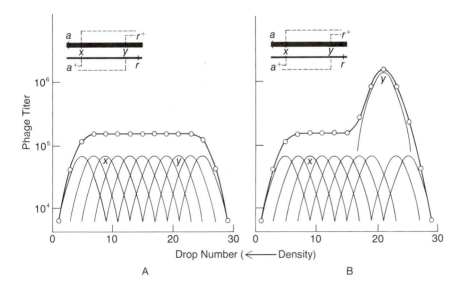

FIGURE 4-2
The expected density distribution of a^+r^+ recombinant phage from replication-blocked crosses. The a parent is density-labeled heavy; the r parent light. The resulting phage lysate is centifuged to equilibrium in a cesium salt density gradient. The expected distribution of a^+r^+ recombinant phage is calculated on the assumption that the product of an exchange event has a simple break-and-join structure. Thus, recombination events that occur at locus x, at 20 percent of the length of the chromosome from the left end, yield a^+r^+ recombinants that are 20 percent light and 80 percent heavy and band at the indicated part of the gradient, centered about drop 9. (The phage titer and drop number scales are arbitrary. The distribution of phage corresponding to a single density species is taken from a typical gradient curve.) The complete distribution of a^+r^+ recombinants is found by adding the contributions from events located (arbitrarily) at 10 percent intervals along the entire length of the chromosome. Graph A shows the expected results corresponding to a uniform distribution of recombination events along the chromosome. Graph B shows the expected density distribution of a^+r^+ recombinants when there is a recombination hot spot at locus y.
(K. D. McMilin, M. M. Stahl, and F. W. Stahl [1974] G 77:409.)

primarily to the RecF pathway, described in Chapter 9.) In Figure 4-3 we see the results of action of the Rec system on unreplicated λ chromosomes in a *gam⁺ red int* infection (similar data are obtained in a *recB* host cell). The following features of the distributions should be noted. (1) Recombinant phage particles (of a^+r^+ genotype, ignoring the C marker for now) have densities from fully heavy (heavy throughout both chains of duplex DNA) to fully light. Furthermore, all intermediate densities are more or less uniformly represented, as expected for a generalized recombination system

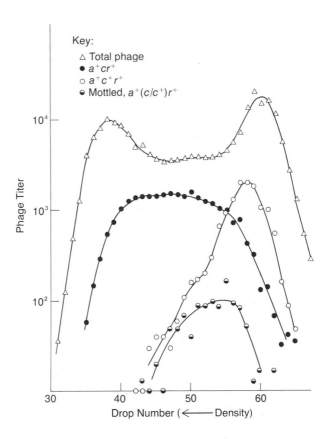

FIGURE 4-3
The density distributions of phage λ produced by the *gam*-inhibited Rec system in a replication-blocked cross between parental types *acr*+ and *a*+*c*+*r*. (F. W. Stahl, K. D. McMilin, M. M. Stahl, J. M. Crasemann, and S. Lam [1974] *G* 77:395.)

operating in such a way that most of the chromosomes have had but one exchange. (2) The curve for the total progeny is composed of three parts. The heavy peak and the light peak arise from exchanges between, respectively, two heavy and two light chromosomes. The intermediate region is composed of equal numbers of the a^+r^+ genotype and, presumably, of the *ar* genotype. (3) a^+r^+ particles are about 25 percent of the total, as expected from both the requirement for dimer formation in the encapsidation process and the random selection of mating partners.

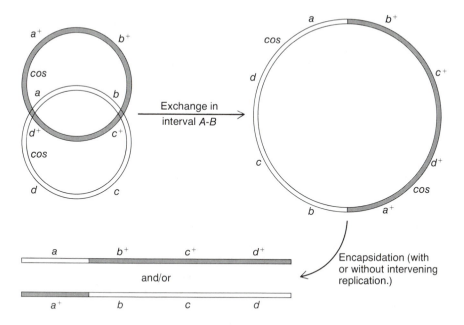

FIGURE 4-4

Exchange between circles yields a linear linkage map in phage λ. Exchange between circular molecules yields dimeric chromosomes. For the conceptually simple example of reciprocal break-join exchange between circles, the product is a dimeric circle. The encapsidation apparatus of λ cuts dimers at *cos* and packages the resulting linear chromosomes into virions.

The assumption that single exchanges predominate is supported by the behavior of the C marker located between A and R. With single exchanges predominating, most of the relatively light a^+r^+ particles should carry the c^+ allele while the heavy a^+r^+ should be c. And so they do.

Above, we have discussed recombination in λ as if the exchanges occurred between linear chromosomes. However, we also made use of the circularity of intracellular λ in order to account for the formation of dimers by exchange, a prerequisite to encapsidation. Are these two views compatible? They are, and Figure 4-4 shows how.

Armed with the simple picture of Rec action, let us now tackle the situation that is normal for λ, the $red^+rec^+gam^+$ one. Results of a replication-blocked cross between heavy and light parents is shown in Figure 4-5. The result is disturbingly different from that observed with Rec alone. The density distribution indicates that exchanges are most frequent near the right end,

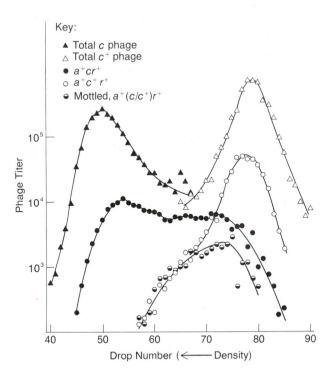

Key:

▲ Total *c* phage
△ Total *c*⁺ phage
● *a*⁺*cr*⁺
○ *a*⁺*c*⁺*r*⁺
◓ Mottled, *a*⁺(*c/c*⁺)*r*⁺

FIGURE 4-5

The density distribution of λ phages produced by the combined
Red and *gam*-inhibited Rec system in a replication-blocked cross
between density-labeled parents of types *acr*⁺ and *a*⁺*c*⁺*r*.
(F. W. Stahl, K. D. McMilin, M. M. Stahl, J. M. Crasemann,
and S. Lam [1974] *G* **77**: 395.)

less frequent near the left end, and least frequent in the middle of the λ
chromosome. That result contrasts with the uniform exchange distribution
seen with Rec alone, and is apparently paradoxical in its own right. The
linkage map of λ resulting from standard crosses (Figure 2-7) is approximately congruent with the chromosome of λ as revealed from electron
microscopy of deletion and substitution mutations (Chapter 9).[26] Thus, in
standard crosses with all recombination systems wild type, the exchange
distribution upon the chromosome must be almost uniform. In replication-
blocked crosses, however, there is a relative paucity of centrally located
exchanges. The conclusions from microscopy and replication-blocked
crosses are reconciled by crosses that permit a little bit of replication. In

such experiments both parents are density labeled in order that the amount of new DNA in any progeny chromosome can be assessed by its density.

The results of one such cross involving two intervals defined by three marked loci are shown in Figure 4-6. One interval is centrally located; the

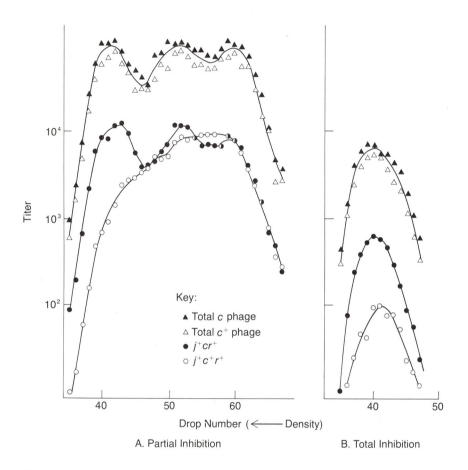

Key:

▲ Total c phage
△ Total c^+ phage
● j^+cr^+
○ $j^+c^+r^+$

Drop Number (⟵ Density)

A. Partial Inhibition   B. Total Inhibition

FIGURE 4-6

The production of recombinants in the middle of the λ chromosome is relatively dependent on replication. Two density-labeled parental types, jcr^+ and j^+c^+r, infected bacteria in which replication was partially or totally inhibited. Among progeny phage bearing unreplicated chromosomes (all phage in graph B and those in the left-most peak in graph A), recombinants in the centrally located $J–cI$ interval are less frequent than those in the shorter, terminally located $cI–R$ interval. Among phage that bear replicated chromosomes (middle and right peak in graph A), recombinants in the $J–cI$ interval are better represented. (F. W. Stahl, K. D. McMilin, M. M. Stahl, and Y. Nozu [1972] *PNAS* **69**:3598.)

other is at the right end. The amount of replication was controlled by varying the temperature using a host cell with a temperature-sensitive DNA replication function. In Figure 4-6B a high temperature was used, and all the emerging phages are fully heavy. The central interval, which is the longer of the two in the standard map and on the chromosome, gives the lower frequency of recombinants. In Figure 4-6A a slightly lower temperature was used, and the appearance of half-heavy and fully light progeny signals replication. Among the fully heavy progeny, the central interval shows the same recombinant shortage; among the fully light progeny, however, the recombinant frequencies in the two intervals reflect more nearly their relative physical lengths. We conclude that generalized recombination in the middle of the λ chromosome fares relatively badly among chromosomes that fail to replicate. Not only does Red facilitate replication, but replication seems to be required for the normal operation of Red as well. A dependence of Red upon replication is apparent also in crosses lacking the Rec system due to a *recA* mutation.[27]

In the absence of replication, Red by itself acts very poorly, judging from the low yields of phage.[28] Furthermore, Red does not seem to act uniformly; the only recombinants found have undergone an exchange near the right end of the chromosome. These conclusions regarding Red are supported by replication-blocked crosses in which both Red and Int are wild type. The Int-mediated recombinants are equal to or in excess of the Red recombinants, whereas, in standard crosses with free DNA replication, Red is more active than Int. Is the low activity of Red due to a shortage of Red enzymes under the replication-blocked conditions or the inability of Red to recombine circular molecules in the absence of replication? Two observations argue for the latter view. (1) Red recombination is poor under conditions in which the Int genes, which are cotranscribed with the Red genes, mediate a high level of recombination.[29] (2) The presence of Red enzymes in replication-blocked crosses is revealed by the high rate of terminal Red recombination in *recA*[+] host cells, described above.

When Red and *recA* are both wild type, recombinant production in replication-blocked crosses is greater than the sum of the separate productions when one or the other is genetically inactivated.[30] Thus, Red's ability to act on unreplicated DNA is aided by *recA*[+], although the activity of Red is still confined to the terminal regions of the chromosome. The replication-blocked crosses described above were all conducted with *gam*[+] phage. Parallel crosses with *gam* phage indicate that the relatively high rate of terminal recombination is partially dependent on inhibition of the *recBC* nuclease. Efforts to rationalize the relations described take us so speculatively into models for mechanisms of exchange that we must postpone the matter until Chapter 10. By then we shall have encountered other features of λ recombination to be accounted for as well.

In this chapter, as in Chapter 2, we took up λ first because of its relative simplicity. Let us now look at T4 single bursts and the way in which T4 recombination fits into its chromosome cycle.

RECOMBINATION AND THE CHROMOSOME CYCLE IN T4[31]

The T4 chromosome in the virion is a Watson-Crick duplex. As mentioned in Chapter 2, a clone of T4 consists of particles of the same genotype but different gene sequence. Each of the particles has a different circular permutation of the 200 or so genes of T4.[32] The chromosomes of T4 have a second important structural feature. Three or four of the genes present at one end of any given particle are repeated at the other.[33] As far as we know, a chromosome can have ends anywhere along its 2×10^5 base pairs.

Unlike λ, T4 supplies all or most of its own replication enzymes. Following transcription and translation of the "early" genes that inform those enzymes, replication is initiated at several points in the infecting chromosome. Successive rounds of replication create a "pool" of linear chromosomes whose sequence is the same as that of the infecting chromosome(s). DNA structures *longer* than T4 then accumulate. We shall see that the mechanism giving rise to these concatemers of T4 holds a central place in our picture of recombination in T4. As with λ, the process of encapsidation reduces the concatemers to monomers. In T4 (unlike λ) encapsidation is insensitive to DNA sequence. Any linear segment of DNA is wrapped up until no more of the molecule will fit into the head. The protruding end is then cut off, and the head is "stoppered." The amount of DNA that fits into a normal T4 head is just a few thousand base pairs longer than the T4 genome, assuring a terminal redundancy in each packaged chromosome. The lack of sequence specificity in packaging cyclically permutes the chromosome among the members of a T4 clone.

The rate of recombination per unit physical length is high in T4. This fact, plus the permuted character of the chromosomes, has rendered density-transfer experiments like those described for λ impossible. (On the other hand, the high rate of recombination has facilitated studies on the state of the intracellular DNA during the recombination process. We shall examine those studies in later chapters.) Recombination analysis in T4 has been further hampered by the lack of mutations that eliminate recombination without being simultaneously lethal. This situation is now understood to reflect a fact of T4 life; recombination plays an absolutely essential role (or roles) in the T4 chromosome cycle.

The use of single bursts for genetic analysis was first applied in T2, a close relative of T4.[34] Single bursts showed little correlation in the frequencies of complementary recombinant types, suggesting nonreciprocality of the exchange process. They showed some, but not much, nonrandomness in the distribution of recombinants of a given type, even for markers with sufficiently small R that most bursts contained no recombinants of that type at all.[35] This result argued that recombination in T2 is frequently, though not always, a terminal event, and is rarely, if ever, an early event.

The terminal redundancy of T4 provides the physical basis for a kind of heterozygosity (see Chapter 5 for a discussion of other kinds). Properties of the heterozygous particles help elucidate the role of recombination in the T4 chromosome cycle and, as a bonus, offer an important clue to the chemistry of T4 exchange.

In a genetically mixed infection, i.e., a cross, some of the concatamers formed are genetically mixed. As we trace along a concatemer (conceptually), a given marked gene may be found in first one allelic state and then another. (In a few moments, we shall explain how the mixed concatemer arises.) In the event that encapsidation packages both copies of the gene into the same particle, a terminal-redundancy heterozygote is formed. In a cross of, say, $ab^+ \times a^+b$ (where A and B are closely linked) a terminal-redundancy multiple heterozygote may result, as for example:

Single bursts of cells infected by such heterozygotes reveal the rules governing transmission of information on the terminal redundancy. Despite the close linkage of A and B in standard crosses, many or even most of the *progeny* of an individual terminal-redundancy heterozygote are recombinant for A and B. This fact (and see Chapter 6) led to the related views that recombination occurs at an extraordinary rate near the ends of T4 chromosomes, and that this head-to-tail exchange (Figure 4-7) between members of an intracellular T4 clone is important in building concatemers.

The role of chromosome ends in T4 recombination is supported by the results of single-burst crosses between normal and petite T4 particles, particles whose head volumes are about two-thirds normal size.[36] These petite particles are "dead" under conditions of single infection but can participate well in infections made jointly with other particles, either petite or normal. In a joint infection the extent of the chromosome fragment in

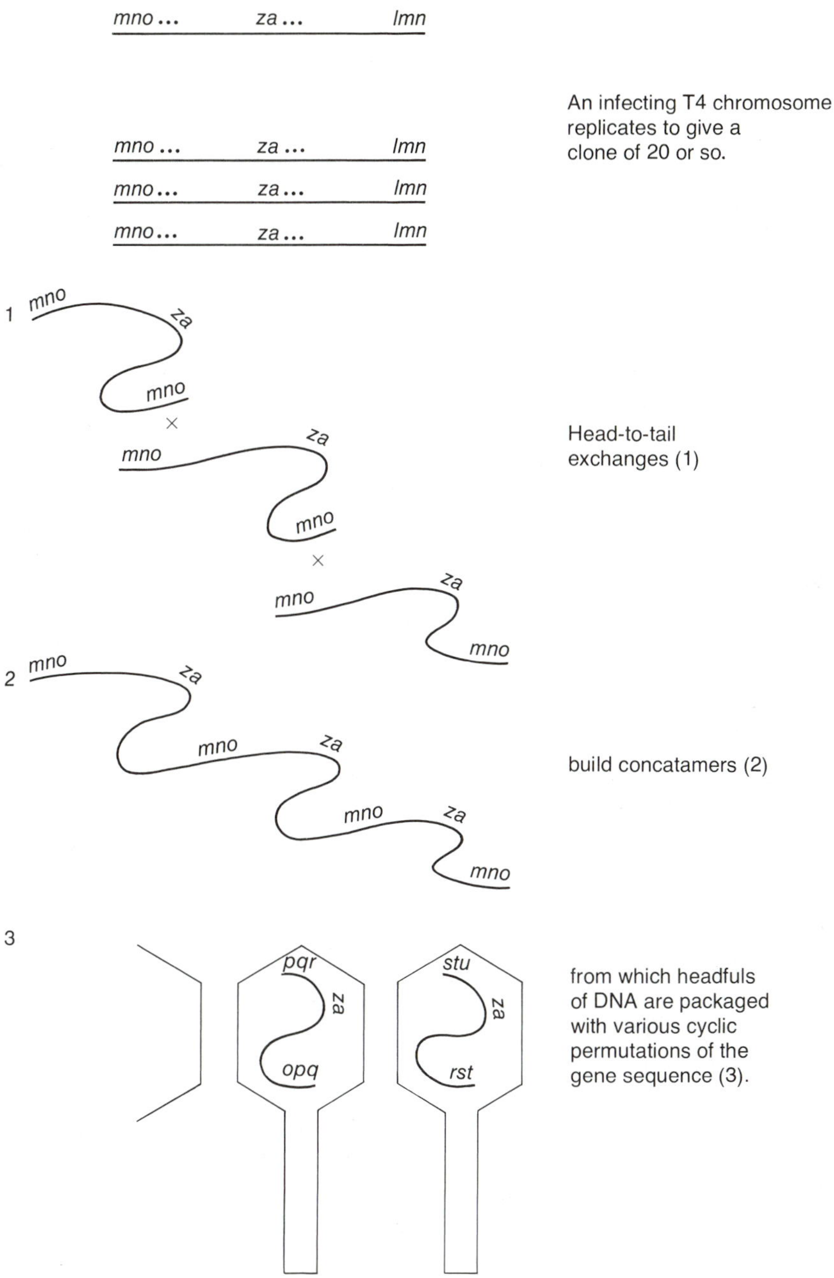

mno ...   za ...   lmn

—

mno ...   za ...   lmn   An infecting T4 chromosome
                                             replicates to give a
mno...   za...   lmn   clone of 20 or so.

mno...   za...   lmn

1 mno   za

     mno
       ×   za
   mno   Head-to-tail
                  mno   exchanges (1)
                    ×
            mno   za
                              mno

2 mno   za

          mno   za   build concatamers (2)

                 mno   za
                            mno

3

              pqr   stu   from which headfuls
                 za   za   of DNA are packaged
                                             with various cyclic
              opq   rst   permutations of the
                                             gene sequence (3).

FIGURE 4-7
Origin of the cyclic permutation of the T4 chromosome.

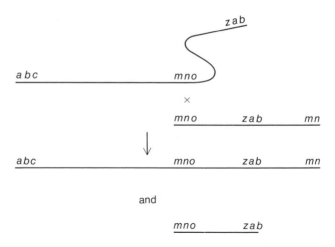

FIGURE 4-8
An exchange between T4 chromosomes from different clones,
occurring at a chromosome end, has a large product from which
a (recombinant) chromosome can be encapsidated.

individual petite particles can be determined by noting the contributions of
markers from petite particles to single bursts. (As expected, all conceivable
chromosome segments two-thirds the length of the T4 genome are found
among such particles.) Since the ends of a single petite chromosome can be
deduced from its genetic contribution, the rate of exchange on that chromosome as a function of distance from the ends can be assessed. Of course the
exchange is not a result of incestuous head-to-tail encounters since a petite
particle has no terminal redundancy. The recombination rate is seen to rise
dramatically at each of the two ends.[37] The result also explains how genetically mixed concatemers arise; a chromosome end can exchange with a
homologous region of a chromosome from a different clone in the same
cell (Figure 4-8).

The role of exchange in concatemer formation is supported by studies on
intracellular T4 "recombinant DNA", i.e., nonencapsidated DNA molecules
that are composed partly of atoms from one parent and partly of atoms
from another or from a daughter molecule.[38] (Density labels and radio-
isotopic labels generally distinguish the two parents in these experiments.)
These molecules, which physically manifest exchanges, are not found (in
normal infections) during the time when the pool of unit-length monomers
is accumulating, but are found later, as concatemers appear. The role of
exchange in concatemer formation is further confirmed by studies on
conditional-lethal mutants that reduce recombination under semipermissive
conditions (Chapter 9). Under nonpermissive conditions such mutants fail
to form both "recombinant DNA" and concatemers!

The requirement for exchange in the T4 life cycle may be twofold. First, encapsidation in T4, as in λ, involves cutting and packaging of chromosome-size DNA pieces from concatemers. If cutting and packaging are part of a unit process, as they are in λ, then failure to produce concatemers by exchange would prevent encapsidation. Second, T4 DNA synthesis appears to depend on recombination. If concatemers fail to accumulate at the normal time, then the rate of DNA replication progressively declines. We shall discuss the importance of concatemers in replication in Chapter 10.

The view outlined above* provides adequate explanation for the apparent nonreciprocality of T4 recombination. An exchange occurring between the end of one molecule and its homologous sequence in the middle of another (Figure 4-8) will produce a maturable recombinant component of a concatemer as one product and a fragment as the other. The small average size of recombinant clones is likewise attributable to features of recombination identified above. First, the enzymes for exchanges are either absent or ineffective until appreciable replication has occurred. Second, since recombination generates concatemers, the substrates for encapsidation, recombinants are apt to be snapped up by hungry heads depriving them of the chance to replicate.

The involvement of chromosome ends in T4 recombination can explain the dependence of r on multiplicity of infection beyond that explained by finite input (Chapter 2).[39] More infecting particles introduce ends in more places along the chromosome, conferring upon those regions a higher recombination rate.

PROBLEMS

4-1 The frequency of heterozygosity due to terminal redundancy rises in the intracellular T4 particle population as a function of time and then levels off. The leveling indicates that equilibrium has been effectively established; further time does not make for better mixing of genes separated by the length of a T4 chromosome. Because g (Chapter 2) is close to unity and because most of the particles in a normal progeny are encapsidated from well-mixed concatemers, the frequency of redundancy heterozygosity is a fair measure of the length of the reduncy itself.

(a) Infection by T4 was made by large equal numbers of a and a^+ particles. In the progeny the frequency of redundancy heterozygotes was 1 percent.

---

* For a different view see G. Mosig (1962) *ZV*, **93**:280.

What is the approximate length of the redundancy as a percent of T4 chromosome length? As base pairs?

(b) If one of the infecting types had been nine times more frequent than the other, what frequency of redundancy heterozygotes would have been observed?

4-2 In T4 strains bearing deletions, redundancies are longer than normal in accord with the size of the deletion from the chromosome.

(a) A mixed infection of a and a^+ was made in which both parents carry another mutation, b_{del}, which deletes about 4 percent of their DNA. What fraction of the progeny will be a/a^+ redundancy heterozygotes?

(b) A mixed infection was made between a deletion mutant missing 4 percent of its chromosome and wild type. What will be the frequency of redundancy heterozygotes bearing the deletion at one end and wild type at the other?

4-3 In λ a cross was performed in which one parent type was density labeled (^{15}N and ^{13}C) while the other was composed of ordinary isotopes. The multiplicities of the two types were equal. Replication was tightly blocked by appropriate mutations in the phage and the host. The cross was $a^+r\,(^{15}N, ^{13}C) \times ar^+\,(^{14}N, ^{12}C)$, where A and R are terminal genes. The parents were *red* and the host was Rec$^-$, so there was no generalized recombination. The Int system however, which operates on *att* about 57 percent of the physical distance from A to R, was functional. The cross yielded four density species of progeny particles, which were displayed like this on a cesium density gradient:

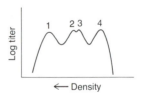

(a) What are the genotypes of phages in each of the four peaks?

(b) Explain how the equality in heights of the four peaks indicates that all mature phages enjoyed an Int-mediated exchange.

(c) If the light parent had been adsorbed to the bacteria at a multiplicity twice that of the heavy one, what would have been the relative heights of peaks 1–4?

4-4 Single-burst crosses with *red int* λ were performed so that the (gamma-inhibited) Rec system could be examined for reciprocality (P. V. Sarthy and M. Meselson (1976) *PNAS* **73**:4613). The cross was $hcmi \times h^+c^+mi^+$. The data from the genetically mixed bursts are in Table 4-1. (They reveal an excess of the *hcmi* parent among the progeny, but this is inconsequential. It is presumably a result of an accidental excess of *hcmi* over $h^+c^+mi^+$ in the infecting phage mixture.)

TABLE 4-1
Data from two single-burst experiments with λ in which recombination is attributable
to the gamma-inhibited Rec system of *E. coli*

	$h^+c^+mi^+$	$hcmi$	hc^+mi^+	h^+cmi	$hcmi^+$	h^+c^+mi	hc^+mi	h^+cmi^+
Rec I	60	64	0	0	0	0	0	0
	57	124	3	4	0	0	0	1
	43	151	3	5	3	4	1	1
	68	143	3	6	0	0	0	0
	47	154	0	0	5	2	1	1
	13	204	0	0	0	0	0	0
	101	160	5	2	2	3	0	1
	46	129	0	0	0	0	0	0
	147	172	2	1	5	0	2	1
	48	264	0	0	0	0	0	0
	96	82	4	5	4	1	2	1
	73	87	3	3	0	0	0	0
	72	181	0	0	4	2	1	1
	200	29	3	2	2	2	1	0
	79	109	3	2	0	0	0	0
	93	183	2	4	0	0	2	1
	126	102	0	0	3	3	0	0
	60	136	0	0	0	0	0	0
	51	64	0	0	0	0	0	0
	56	46	0	0	0	0	0	0
	77	48	0	0	2	2	1	1
	83	129	6	3	5	4	1	0
	58	72	0	0	0	0	0	0
	47	92	0	0	2	3	1	0
	102	92	3	3	3	3	1	1
	92	48	0	0	0	0	0	0
	64	95	3	2	0	0	1	0
	70	92	0	0	1	2	1	1
	90	77	3	3	3	5	1	1
	93	126	0	3	3	6	0	1
	72	107	0	0	3	3	2	2
	67	77	0	0	1	0	0	0
	60	155	4	5	0	0	1	2
	36	84	0	0	0	0	0	0
	98	135	3	1	2	2	3	0
Subtotal	2645	4013	53	54	53	47	23	17
Rec II	61	42	2	3	0	0	0	0
	82	90	2	4	6	3	1	1
	25	83	2	3	0	0	0	0
	104	137	0	3	4	2	0	0
	30	97	1	3	1	1	0	0
	35	100	4	2	0	0	0	1
	78	126	0	0	1	1	0	0
	162	152	2	2	10	4	0	0
	103	167	5	8	0	0	0	0
	76	62	5	6	2	2	0	0
	43	151	0	0	0	0	0	0

(*continued*)

TABLE 4-1 (*continued*)

	$h^+c^+mi^+$	$hcmi$	hc^+mi^+	h^+cmi	$hcmi^+$	h^+c^+mi	hc^+mi	h^+cmi^+
	68	160	0	2	3	3	0	0
	48	78	0	0	6	4	0	0
	93	113	8	11	0	0	0	0
	86	56	3	2	2	4	0	1
	27	36	0	0	0	0	0	0
	37	164	3	2	0	2	1	1
	42	112	1	4	3	2	0	1
	87	108	3	8	4	1	0	0
	9	131	0	0	0	0	0	0
	27	112	0	0	0	0	0	0
	45	96	0	0	0	0	0	0
	109	119	3	3	0	0	0	0
	5	77	0	0	2	3	0	0
	115	92	2	4	0	0	0	0
	60	67	2	4	6	3	1	0
	93	131	3	5	1	0	0	2
	81	131	2	5	0	0	0	0
	78	131	3	3	3	4	0	2
	58	104	0	0	4	2	0	0
	70	159	2	4	1	6	2	0
	72	90	4	2	1	1	0	1
	4	107	0	0	0	0	0	0
	86	114	2	4	3	2	0	0
	128	175	4	1	0	0	0	0
	41	111	2	3	0	0	0	0
	10	152	0	1	1	0	0	0
	85	138	5	4	3	2	0	1
	70	118	5	6	3	2	0	2
	83	120	2	7	1	2	0	0
	70	127	0	0	0	0	0	0
	73	184	2	5	0	6	1	0
	65	132	4	4	0	0	0	0
	52	114	0	0	1	3	0	1
	77	142	0	0	3	2	0	0
	87	177	7	6	1	1	0	1
Subtotal	3040	5385	95	134	76	68	6	15
Total	5686	9398	148	188	129	115	29	32
Mean	70.2	116	1.83	2.32	1.59	1.42	0.36	0.40
Variance	1117	1674	3.72	5.77	3.92	2.78	0.44	0.36

Source: Sarthy and Meselson, 1976.

(a) Of the 81 total bursts, how many have the recombinant hc^+mi^+? How many have the complementary recombinant, h^+cmi? If the presence of hc^+mi^+ and h^+cmi were uncorrelated, how many bursts would have both types? How many would have one type without the other? Compare your answers with the numbers observed.

(b) How many bursts have the type $hcmi^+$, which is not complementary to hc^+mi^+? If the presence of these two types were uncorrelated, how many bursts would you expect to have both types? How many would have one type but not the other? Compare your answers with the numbers observed.

4-5 The single-burst data for a red^+ gam^+ cross in $recA$ host cells are in Table 4-2. Calculate the expectations for no correlation between the presence of the types hc^+mi^+ and h^+cmi, on the one hand, and for hc^+mi^+ and $hcmi^+$, on the other. Compare them with the observations.

TABLE 4-2
Data from two single-burst experiments with λ in which recombination is attributable to the Red system of λ.

	$+++$	$hcmi$	$h++$	$+cmi$	hc^+	$++mi$	h^+mi	$+c^+$
Red I	74	114	7	2	3	0	0	0
	40	35	0	0	0	0	0	0
	156	16	2	0	6	3	0	0
	76	45	4	0	4	0	1	2
	85	51	4	27	6	10	1	10
	106	2	0	0	0	0	0	0
	3	49	12	1	0	0	0	0
	47	6	0	3	1	0	0	0
	38	15	0	9	3	4	0	1
	30	24	2	2	10	10	1	2
	120	92	0	4	3	2	0	2
	70	106	5	3	2	6	2	2
	132	130	2	4	4	4	1	0
	140	8	0	0	2	0	0	0
	132	10	4	3	2	0	0	0
	185	35	3	6	4	2	0	2
	140	8	2	31	0	0	1	1
	44	122	3	12	2	0	0	1
	109	36	11	4	13	33	3	8
	40	135	0	54	0	4	0	3
	42	26	0	3	11	0	0	7
	89	214	21	13	17	3	7	4
	72	40	4	0	4	0	2	0
	57	25	0	2	1	3	0	0
	4	128	0	2	1	3	0	0
	59	7	1	1	4	1	0	0
	44	48	4	7	5	12	2	2
	4	85	0	3	0	0	0	0
	56	72	0	26	0	2	0	0
	54	30	6	5	4	0	0	1
Subtotal	2248	1714	97	227	112	102	21	48
Red II	91	145	10	15	27	13	1	10
	73	34	1	0	1	2	0	0
	112	4	0	0	1	0	0	0

(continued)

TABLE 4-2 (*continued*)

	$+++$	$hcmi$	$h++$	$+cmi$	$hc+$	$++mi$	$h+mi$	$+c+$
	154	24	5	12	6	2	0	1
	148	32	3	0	1	29	0	4
	100	56	10	8	3	0	0	0
	165	88	13	5	6	3	0	0
	83	61	4	2	7	3	0	1
	108	13	0	14	6	4	0	5
	77	64	8	1	13	0	0	1
	44	80	1	1	7	0	2	2
	26	26	0	13	2	2	0	0
	123	4	3	1	4	0	0	0
	69	34	3	6	6	1	0	5
	106	38	2	2	0	0	0	0
	132	109	7	0	28	2	0	3
	134	15	7	4	4	1	1	5
	130	78	0	5	1	0	0	0
	142	8	0	1	10	0	0	1
	79	3	4	6	5	3	0	0
	78	64	8	7	13	3	1	1
	109	79	8	5	10	4	0	0
	52	57	32	12	19	0	1	8
	139	78	7	6	5	0	0	1
	89	30	5	7	12	0	0	4
	17	115	16	2	0	15	6	0
	144	25	12	3	16	1	0	3
	103	22	9	5	8	1	0	2
	66	145	0	0	3	0	0	0
	101	33	0	12	0	0	0	0
	175	46	0	0	10	0	0	0
	180	38	0	0	1	0	0	0
	137	63	3	7	10	0	0	1
	115	23	2	11	1	1	0	0
	175	24	0	0	0	0	0	0
	131	10	10	5	0	0	0	1
	74	54	0	4	4	1	0	0
Subtotal	3981	1822	193	182	250	91	12	59
Total	6229	3536	290	409	362	193	33	107
Mean	93.0	52.8	4.33	6.10	5.40	2.88	0.49	1.60
Variance	2119	1914	28.3	75.0	46.4	34.9	1.54	5.96

Source: Sarthy and Meselson, 1976.

4-6 Write a mapping function for λ assuming $g = 0.4$. Solve it for m, letting $r = 0.15$ at $R = 0.5$. Graph r versus x and compare it with Figure 2-6, which shows the λ mapping function at $g = 0.8$ and $m = 1.0$. Since r values of 0.15 are observed in λ crosses, what is the smallest possible value for g that could apply? What number of participating phages does this represent?

Phage Replication and Single-Burst Analysis

NOTES

1. C. V. Bresch (1955) *ZN* **10b**:545.

2. S. E. Luria (1951) *CSHSQB* **16**:463.

3. C. Steinberg and F. Stahl (1961) *JTB* **52**:488.

4. L. W. Enquist and A. Skalka (1973) *JMB* **75**:185.

5. A. Skalka, M. Poonian, and P. Bartl (1972) *JMB* **64**:541; R. G. Wake, A. D. Kaiser, and R. B. Inman (1972) *JMB* **64**:519; S. Takahashi (1974) *BBRC* **61**:657; D. Bastia, N. Sueoka, and E. C. Cox (1975) *JMB* **98**:305.

6. F. W. Stahl, K. D. McMilin, M. M. Stahl, R. E. Malone, Y. Nozu, and V. E. A. Russo (1972) *JMB* **68**:57; M. Feiss and T. Margulies (1973) *MGG* **127**:285; D. Freifelder, L. Chud, and E. E. Levine (1974) *JMB* **83**: 503; B. Hohn (1975) *JMB* **98**: 93; M. Syvanen (1975) *JMB* **91**: 165; L. W. Enquist and A. Skalka (1973) *JMB* **75**: 185.

7. H. Murialdo (1974) *V* **60**:128.

8. E. L. Wollman and F. Jacob (1954) *AIP* **87**:674.

9. E. R. Signer (1971) in *The Bacteriophage Lambda*, A. D. Hershey, ed., Cold Spring Harbor Laboratory, p. 139.

10. A. J. Clark and A. D. Margulies (1965) *PNAS* **53**:451.

11. T. Tomizawa and H. Ogawa (1972) *NNB* **239**:14.

12. A. J. Clark (1971) *ARM* **25**:437.

13. L. W. Enquist and A. Skalka (1973) *JMB* **75**:185.

14. A. Skalka (1974) in *Mechanisms in Recombination*, R. F. Grell, ed., New York: Plenum, p. 421.

15. L. W. Enquist and A. Skalka (1973) *JMB* **75**:185.

16. J. Zissler, E. Signer, and F. Schaefer (1971) in *The Bacteriophage Lambda*, A. D. Hershey, ed., Cold Spring Harbor Laboratory, p. 455.

17. A. Folkmanis and D. Freifelder (1972) *JMB* **65**:63.

18. D. Freifelder, N. Baran, L. Chud, A. Folkmanis, and E. E. Levine (1975) *JMB* **91**:401.

19. J. Weil (1969) *JMB* **43**:351.

20. PV. Sarthy and M. Meselson (1976) *PNAS* **73**:4613.

21. A. Wilkins and J. Mistry (1974) *MGG* **129**:275.

22. D. G. Ross and D. Freifelder (1976) *V* **74**:414.

23. J. Weil (1969) *JMB* **43**:351.

24. W. Fangman and M. Feiss (1969) *JMB* **44**:103.

25. K. D. McMilin and V. E. A. Russo (1972) *JMB* **68**:59; F. W. Stahl, K. D. McMilin, M. M. Stahl, and Y. Nuzo (1972) *PNAS* **69**:3598.

26. A. Campbell (1971) in *The Bacteriophage Lambda*, A. D. Hershey, ed., Cold Spring Harbor Laboratory, p. 13.

27. F. W. Stahl, K. D. McMilin, M. M. Stahl, and Y. Nuzo (1972) *PNAS* **69**:3598.

28. F. W. Stahl, K. D. McMilin, M. M. Stahl, J. M. Crasemann and S. Lam (1974) *G* **77**:395.

29. Ibid.

30. Ibid.

31. T. R. Broker and A. H. Doermann (1975) *ARG* **9**:213; F. W. Stahl (1968) in *Replication and Recombination of Genetic Material*, Canberra: Australian Academy of Science, p. 206.

32. G. Streisinger, R. S. Edgar, and G. H. Denhardt (1964) *PNAS* **51**:775; C. A. Thomas Jr. and I. Rubenstein (1964) *BJ* **4**:93.

33. J. Sechaud, G. Streisinger, J. Emrich, J. Newton, H. Lanford, and M. M. Stahl (1965) *PNAS* **54**:1333; G. Streisinger, J. Emrich, and M. M. Stahl (1967) *PNAS* **57**:292.

34. A. D. Hershey and R. Rotman (1949) *G* **34**:44.

35. F. W. Stahl (1956) *V* **2**:206.

36. G. Mosig, R. Ehring, W. Schliewen and S. Bock (1971) *MGG* **113**:51.

37. A. H. Doermann and D. H. Parma (1967) *JCP* **70** (Suppl. 1):147.

38. J. Tomizawa (1967) *JCP* **70** (Suppl. 1):201; T. R. Broker and A. H. Doermann (1975) *ARG* **9**:213.

39. G. Mosig (1962) *ZV* **93**:280.

5

Heteroduplexes

Products of a phage cross and meiotic products in fungi sometimes produce two kinds of progeny. These "heterozygotes" have their physical basis in the duplex structure of DNA and so are called heteroduplexes. The properties of heteroduplexes indicate that in a short region (10^3 or so base pairs) the complementary chains are derived from two recombining duplexes. In some cases heteroduplex formation signals chromosome exchange, as revealed by the recombination of flanking markers. In other cases the maintenance of a parental arrangement of flanking markers suggests that exchange did not accompany heteroduplex formation. The possibility of an even number of very close exchanges is difficult to exclude in the latter case. It seems likely that heteroduplexes signal the mechanism by which chromatids of fungi and chromosomes of phages achieve precise alignment during genetic exchange.

HETERODUPLEXES IN LAMBDA

In the previous chapters we diagrammed chromosomes as lines, ignoring the fact that they are DNA duplexes composed of complementary chains wound about each other. In these simplified, linear diagrams an exchange is a point of union of chromosome segments whose information is derived from two different chromosomes. Beginning with this chapter we acknowledge the duplex structure of chromosomes and find that an exchange has not one but two dimensions, and that our vocabulary must expand accordingly.

Figure 4-3 shows the density distribution of recombinant chromosomes from a density-labeled, replication-blocked λ cross. Note in that figure that in addition to the denser recombinants in the *A-cI* interval and the lighter ones in the *cI-R* interval there were recombinant chromosomes of intermediate density (mottled types) that gave plaques containing two kinds of phage, c and c^+. The particles are heterozygotes. Such particles could indicate that the points of exchange on the two chains of the chromosome are different, one to the left and one to the right of the c marker, as in the following diagram (the arrowheads remind us that the two chains in a duplex have opposite chemical polarity):

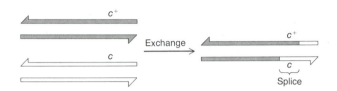

Before we argue for this structure, let us tidy up some definitions relevant to it. A chromosome exchange involves two sites of exchange, one on each chain of the chromosome. The region between the two chain exchanges we call a *splice*. If the two halves of a splice carry different information by virtue of a genetic difference between the recombining chromosomes, then the splice is a heteroduplex splice. Three different arguments support the idea of a splice structure for the λ heterozygotes in Figure 4-3. (1) They cannot be heterozygotes of the terminal redundancy sort, as seen in T4, because λ has no terminal redundancy. (2) By all physical tests the two chains of a λ chromosome are covalently continuous, so there seems to be no possibility for structures such as this:

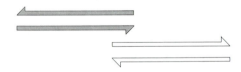

(3) The most beguiling argument is one of economy: when segments of chromosomes are rejoined during an exchange, *something* must ensure their joining in exact register. What better candidate for this job could there be than the formation of a Watson-Crick duplex between homologous stretches of exposed complementary polynucleotide chains? Though you may be less than overwhelmed by these arguments supporting the splice model of chromosome exchange, you will find that the utility of the model in the remainder of the book is justification enough for its adoption. Acceptance of argument 3 leads to a corollary conclusion: chromosome exchanges generally involve splices. Direct proof of this deduction has proven impossible, for reasons that may be apparent later.

The data in Figure 4-3 permit an estimate of the average length of a Rec-mediated splice in λ. If we take the recombinants for what they appear to be, the products of single exchanges, then the length of the splice relative to the A-R interval is simply the fraction of the recombinants that are heterozygous at cI. This fraction is 0.028, which is equivalent to about 10^3 base pairs. An additional assumption in the calculation is that all splices that embrace a cI marker are manifest as heterozygotes. For both trivial and interesting reasons, this is unlikely to be true. The trivial reasons relate to the difficulty of identifying genetically mixed plaques; the interesting one is treated in Chapter 6. Thus, our estimate of splice length is apt to be an under-estimation. If the under-estimation is not too great, we can argue that these splices are about a gene long. Later in this chapter we shall further examine the matter of splice lengths.

HETERODUPLEXES IN FUNGI

If heteroduplexes exist in fungi, they should appear as the progenitors of genetically mixed clones arising from individual haploid products of meiosis, and indeed they have been found as such in all fungi where they have been sought. The splice model anticipates that these heteroduplexes will be recombinant for markers flanking the splice. Hindsight tells us that markers with the following distances are optimal:

$R_{AB} = R_{BC} =$ a few percent

$R_{AC} \cong R_{AB} + R_{BC}$

Genes A and C are far enough from B that heteroduplex splices embracing B will generally not extend as far as A or C. On the other hand, A and C are close enough to B that additional exchanges unrelated to the one that produces the splice on the B marker will rarely occur. Among the (rare) heteroduplexes found at B, as many as half (rarely more) are recombinant for the marked genes A and C. Since the overall rate of recombination for A and C is only 10 percent, there is an evident positive correlation between the heteroduplex state at B and an exchange between A and C. This is pretty good evidence for splices in fungi. What, however, are those other hetero-duplexes—the half or more that are not recombinant for A and C? Two types of explanations seem plausible. Either the heteroduplexes are splices, but *A* and *C* fail to recombine because a second chromosome exchange occurred between them, or heteroduplexes can arise as a result of something other than splices. At this point, therefore, we hypothesize a new structure, a *patch*, and diagram it like this:

A patch, like a splice, involves two points of exchange. But whereas in a splice the points are on different chains, in a patch they are on the *same* chain.

Before discussing the implications of these alternative explanations for heteroduplexes with nonrecombinant (parental) arrangement of flanking markers, let us sound the warning that a clear operational distinction between them may be close to impossible. If the two exchanges in a double exchange occur close together compared to the length of a splice, then some of the products of a double exchange might look like this:

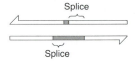

Note that this structure approaches that of a patch as the distance between the splices becomes less. Although an operational distinction between the patch and close double splice may be difficult, they are conceptually distinct.

The data of Figure 4-3, which necessitated the notion of a splice, demonstrate that exchange (via the Rec system, at least) involves the breakage and rejoining of large chromosome segments by splicing; i.e., matter as well as information has been exchanged. For simplicity, let us suppose that the respective contributions of the two interacting chromosomes within the limits of the splice or patch are material as well. In Chapter 11 we shall consider the possibility that this may not always be the case.

The possibility of close double exchanges requires us to alter our thinking about chromosome exchanges and interference. In Chapters 1 and 3 we learned that chromosome exchanges occur randomly along the chromosome in a minority of fungi and actually interfere with each other in most. The double-exchange hypothesis of heteroduplexes with parental arrangement of flanking markers asks us to believe that this is a superficial view. It says that some chromosome exchanges occur as tightly coupled pairs or perhaps even larger clusters,[1] while others may occur singly. (In Chapter 8 we shall see that the members of a supposed cluster show total negative chromatid interference; i.e., they are confined to the same pair of chromatids. This represents another violation of "proper" exchange behavior as we learned it in Chapter 3). As long as the distances between members of a cluster are small compared with the distances between markers, clusters with even numbers of exchanges will go unnoticed, and the recombination we have described in Chapters 1 and 3 will be due only to the clusters with odd numbers of exchanges. These "odd clusters" must generally interfere with each other ($S < 1$); whether an even cluster interferes with an odd or with other evens is not answered by our observations to this point. If we are to think about clustered chromosome exchanges, we clearly have a lot of work ahead of us, and in Chapter 6 we shall start to tackle it.

The patch model has equally painful implications. Acceptance of it means acceptance of a mechanism other than chromosome exchange for the recombination of linked markers. Note that one chain of the duplex chromosome in the patch model is a "double recombinant" whenever the patch falls on a middle marker like this:

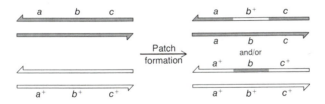

Whatever name we give to the "double recombinant" single chain, it is clear that one round of chromosome replication will provide it with a complementary partner chain, and it will then be a full-fledged double-recombinant chromosome. We shall investigate the probable role of patches on recombination in Chapter 6.

HETERODUPLEXES IN T4

Terminal-redundancy heterozygotes interfere with the detection and study of heteroduplexes in T4. However, the first evidence for the existence of both redundancy heterozygotes and heteroduplexes came from examination of progenies from individual heterozygous particles.[2] As we described in Chapter 4, the offspring of a redundancy heterozygote are highly recombinant; in fact, with multiply marked heterozygotes a stronger statement can be made. Consider the following redundancy heterozygote (with all markers in the *rII* region):

The progeny from such a particle will be highly recombinant for the *r* markers, but reciprocal recombinants will be grossly unequal. In fact, *most* of the particles in a single burst will be $r_1^+ \ldots r_4$ (where the dots mean you should not worry now about the state of markers *2* and *3*) so that the marker r_1^+ greatly outnumbers r_1 and r_4 outnumbers r_4^+. Similarly, r_3 outnumbers r_3^+, but less strongly so. We may find r_2 and r_2^+ equal or even find r_2^+ outnumbering r_2. In each terminal heterozygote, markers tend to get lost according to their closeness to the chromosome ends (polarized segregation.) For clarity, I have inverted the actual argument; for any given heterozygous particle the chromosome structure is in fact deduced from the observed polarized segregation. A heterozygote that produces more r_4^+ progeny particles than r_4 and more r_1 than r_1^+ would be diagrammed like this:

Among the progeny of a standard cross, many multiple heterozygotes manifest polarized segregation and are therefore of the redundancy sort. Among the progeny of such heterozygotes, however, multiple heterozygotes are found that do not show polarized segregation. Their different behavior argues for a different structure. The strong candidate is the heteroduplex, either splice or patch. This result is in accord with the view, presented in Chapter 4, that head-to-tail recombination between terminal redundancies builds up concatemers. The lateness of concatemer formation allows some of the resulting heteroduplexes to mature without an intervening replication.

Heteroduplexes can be demonstrated in T4 more directly. Crosses conducted in the presence of 5-flourouracil deoxyriboside (FUDR) suffer diminished DNA replication but are not blocked for genetic recombination. Since heteroduplexes arise during recombination and are destroyed by replication, they accumulate in FUDR crosses.[3] After sufficient time, encapsidation results in a yield of heterozygotes the majority of which are heteroduplexes. Most of the heteroduplexes retain parental configurations of flanking markers. Within the framework of the patch model, therefore, patches outnumber splices, at least in FUDR crosses. However, in view of the possible clustering of chromosome exchanges under some conditions in T4 (Chapter 6), the high frequency of heteroduplexes with parental flanking markers may well be due in part to coupled splices.

A separate, indirect line of evidence supports the view that patches do occur in T4 at least under some conditions. The argument is so indirect that it starts with studies on genetic transformation in the pneumococcus bacterium.

Transformation involves the introduction of genetically marked *fragments* of DNA into receptive ("competent") cells. A fraction of these cells give rise to clones genotypically like the strain from which the DNA fragments were derived. The interaction of a fragment with a recipient chromosome obviously involves some kind of *even* multiple "exchange" (or double exchange, to be simple). If it were done with splices, we could diagram the reaction this way:

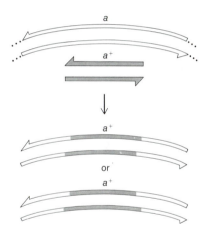

This invites the question, for which we have an answer below, of how the cell is smart enough not to indulge in a lethal, odd number of exchanges.

The double-exchange model is challenged by analyses of DNA removed from recipient cells promptly after transformation with density-labeled DNA.[5] The recovered DNA is fragmented in a controlled stepwise manner, and the fragments from successive steps are sorted out by density. The donor markers are located in the density gradient using transformation as a bioassay. The double-exchange model predicts the consequences of successive fragmentation. With long DNA pieces, the recovered marker should be in fragments whose density is essentially that of the recipient DNA. With successive fragmentation, the marker-bearing pieces should get progressively more dense until they achieve a density equal to the transforming DNA prior to its use in the experiment. Such was not found, however. Successive fragmentation revealed a limit in density halfway between the recipient and the original transforming DNA. Thus, within the framework of the double-exchange model, the splices are longer than the region of donor duplex, and we should modify our diagram of a transformed chromosome to look like this:

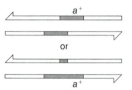

Actually, however, the above experiment failed to reveal any donor fragments of original density, so that economy of hypothesis suggests we set the splice length at zero, giving:

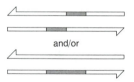

With the splice gone from both diagrams, we have patches. The mechanism of assuring an even number of "exchanges" thus appears to be circumvention of chromosome exchange altogether and the use of patches instead.[6] The discovery that transforming DNA becomes single chained prior to its incorporation into the recipient chromosome further supports this conclusion and leads us to accept the diagram with patches as a better representation of a transformed chromosome. Step two in our chain of evidence for patches in T4 is simply that single-chain DNA fragments are effective transforming agents in that system, too.[7]

The high rate of recombination in T4 (Chapters 2 and 6) has assisted efforts to visualize recombination reactions by electron microscopy.[8] These studies use phage strains genetically blocked for DNA replication. Under these conditions, any branched DNA structure is likely to be a recombinational intermediate. In infections made with a mixture of density-labeled and ordinary T4, the intermediate density of branched structures attests to their biparental origin. Of course, the degree to which the visualized intermediates are artifacts of blocking replication is uncertain. Structures most relevant to splices and patches can be diagrammed like this:

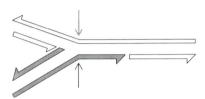

The introduction of a cut at the place indicated by the gray arrow would produce a splice; a cut at the place indicated by the black arrow would result in a patch. The symmetry of the structure suggests that the two modes of resolution may be equally probable. These structures obviously cannot produce recombinants in complementary pairs, a deduction concordant with the apparent lack of reciprocality of T4 recombination as described in Chapter 4.

Estimates of the average length of splice or patch usually fall in the range of 10^3 base pairs, with a wide variability from one estimate to another. Different methods have been applied, and the agreement, though rough, is heartening. An estimate of splatch (splice and/or patch) lengths can be obtained from measurements on structures like those diagrammed above.[9] They give a median value $\geqslant 420$ base pairs. Other methods are "genetic." One of these, described early in this chapter, gave an estimated splice length of 10^3 base pairs for the Rec system operating on λ. Both of the above methods are rather special cases. The most frequently employed method uses close multiple markers. In this method, first the heteroduplexes for one marked site are identified, and then the fraction of these that are heterozygous for a nearby site is determined. This information yields an estimate of average length if one is willing to make several assumptions.[10] (1) A length distribution must be assumed (or determined, if many markers are used.) (2) It must be assumed that the markers do not influence the location or length of the splatch. (3) We must assume that all heterozygotes at the second site are detected. In particular, we must suppose that none are corrected (Chapter 6). (4) Of course, we must know the physical distance between the markers employed, and this knowledge sometimes involves assumptions (see Chapter 10). Despite these uncertainties, various estimates by this method have all placed the average splatch length in the range of 10^2–10^3 base pairs (see Problems 5-1 and 5-2).

PROBLEMS

5-1 Harris made the cross $r_1 r_2 \times r_1^+ r_2^+$. The two sites are separated by a distance D. He identified particles which were heteroduplex at site 1. He thought he was dealing with a mixture of splices and patches but assumed they were of equal, constant length L.

(a) Write an expression in terms of D and L for the fraction of site 1 heteroduplexes that were heteroduplex at site 2.

(b) If Harris found that 30 percent of the site 1 heteroduplexes were heteroduplex for site 2, what was his estimate of L in terms of D?

(c) If Harris showed that r_1 and r_2 caused amino acid replacements that were 50 residues apart on the protein product, what is his estimate for L in base pairs?

5-2 Use the concocted data from Problem 5-1. Assume that splatch lengths are exponentially distributed. Now calculate L, the mean splatch length, in base pairs. (Equation A-3 and Figure A-3, in the appendix, describe the exponential probability distribution.)

5-3 It can be reasoned that splices must have a minimal length greater than a few base pairs if they are the mechanism by which precise exchange is achieved. For instance, a splice one base pair long could incorrectly form in a number of positions equal to one-fourth the number of base pairs in the entire chromosome. Now, assume base sequences are random (which, of course, they are not really).

(a) How many times would a given sequence of two base pairs occur in a chromosome 10^7 base pairs long (the approximate length of an E. coli chromosome or a large fungal chromosome)?

(b) How many times would a given sequence of three base pairs occur in a chromosome of 10^7 base pairs?

(c) How long must a sequence be in order for us to say with greater than 99 percent certainty that it will occur only once in 10^7 base pairs?

5-4 How many distinguishable types of heteroduplexes can arise in the one-factor cross $m^+ \times m$, where the mutant, m, differs from the wild type, m^+, by a single base-pair transition?

5-5 In T4, the frequency of heteroduplexes among the first matured chromosomes is the same as that observed at the time of spontaneous cell lysis. During this same period, recombinant frequencies rise about twofold for close markers (see Figure 2-2). Is this compatible with the view that heteroduplexes signal the region in which an exchange has occurred? How?

5-6 In yeast, heteroduplex ascospores are identified by the genetically mixed colonies to which they give rise. Two heteroduplexes for the same marked site are almost never identified among the four meiotic products of a single ascus. Think about what this tells us about heteroduplex formation in yeast.

Try to reconcile your thoughts with the well-established fact that chromosome exchange in yeast is reciprocal, i.e., that exchange produces complementary recombinants in pairs.

NOTES

1. R. H. Pritchard (1960) *GR* **1**:1.
2. A. H. Doermann and L. Boehner (1963) *V* **21**:551.
3. J. Sechaud, G. Streisinger, J. Emrich, J. Newton, H. Lanford, H. Reinhold, and M. M. Stahl (1965) *PNAS* **54**:1333.
4. H. Berger (1965) *G* **52**:729.
5. M. S. Fox and M. K. Allen (1964) *PNAS* **52**:412; W. F. Bodmer and A. T. Ganesan (1964) *G* **50**:717; N. Notani and S. H. Goodgal (1966) *JGP* **49** (part 2):197; T. Gurney and M. S. Fox (1968) *JMB* **32**:83.
6. But generalized transduction by the *Salmonella* phage *P22* proceeds by chromosome exchanges! (J. Ebel-Tsipsis, M. S. Fox, and D. Botstein (1972) *JMB* **71**:449)
7. G. Veldhuisen and E. B. Goldberg (1968) in "*Methods in Enzymology*," L. Grossman and K. Moldave, eds., Academic Press, New York, p. 858.
8. T. R. Broker (1973) *JMB* **81**:1.
9. Ibid.
10. C. Levinthal (1954) *G* **39**:169.

6

Very Close Markers

Random spore and particle analysis reveals localized negative interference (LNI); for very close markers the frequency of double recombinants in three-factor crosses is higher than the product of the recombination frequencies for the two intervals involved. Three sources of LNI have been identified: (1) patches, the result of paired chain exchanges confined to one chain of a duplex; (2) clusters of exchanges, either splices (chromosome exchanges) or patches, or both; and (3) correction of base mismatches and/or nonmatches in heteroduplex splices and patches. For T4, patches and clusters can account for most observations. For λ, patches and clusters have not been demonstrated, while correction demonstrably contributes to LNI. In fungi, patches appear likely and can go a long way in accounting for LNI. Clusters of chromosome exchanges are almost as good an explanation. However, map expansion indicates marker effects that might signal heteroduplex correction.

LOCALIZED NEGATIVE INTERFERENCE

The rules of recombination stated in our first four chapters must be bent, sometimes to the point of breaking them, in crosses involving very close markers. The definition of "very close" is circular; it is that distance at which the rules get into trouble. Roughly speaking, this distance turns out to be about the length of one or a few genes, i.e., $10^3 - 10^4$ base pairs. In this chapter we shall examine very close marker recombination as it is manifest in random spore and random particle analysis.

The coefficient of coincidence in three-factor crosses in which at least one of the intervals is short always rises as the markers are moved from "close" to "very close." The data for λ (Figure 2-5) illustrate the phenomenon. From a plateau of about 3, s rises as r decreases, reaching a value of 72 in the

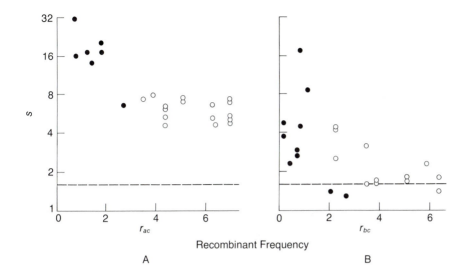

FIGURE 6-1

Relationship of the coefficient of coincidence ("s") to the recombinant frequency (r) for markers within the *rII* region of phage T4. In *A* the crosses were $ab^+c \times a^+bc^+$, where "s" = twice the frequency of $a^+b^+c^+$ divided by the product of the frequencies r_{ab} and r_{bc} measured as twice the wild-type frequencies in the corresponding two-factor crosses. In *B* the crosses were $ab^+c^+d \times a^+bcd^+$, where "s" = twice the frequency of $a^+b^+c^+d^+$ divided by the product of the frequencies r_{ab} and r_{cd} measured as twice the wild-type frequencies in the corresponding two-factor crosses. The filled circles represent values from intracistron crosses and the open circles values from intercistron crosses. The horizontal lines at "s" = 1.6 represent a value of "s" expected simply from mating theory considerations. (M. Chase and A. H. Doermann [1958] *G* **43**:332.)

experiments reported. T4 behaves similarly (Figure 6-1). For fungi, S, which is generally close to zero for close markers, rises above unity for very close ones. This rise in the coefficient of coincidence as markers become very close is called *localized negative interference* (LNI), or sometimes high negative interference.[1]

LNI is a formality. In Chapter 1 we defined interference as a nonrandomness in the distribution of chromosome exchanges and pointed out that nonrandomness would influence the frequency of double recombinants. This influence is conventionally measured as $S = R_{\text{doubles}}/(R_{AB} \times R_{BC})$. If recombinational processes other than chromosome exchange exist, these will contribute to R and may dictate S values. Thus, when we find S rising above unity in crosses with very close markers, we are not necessarily observing the consequences of a nonrandom distribution of chromosome exchanges.

Nevertheless, in this book, LNI will *always* mean that S (or s or "s") rises as R (or r) values become very small. The possible reasons for the rise will concern us in this and the following chapters.

No less than three possible contributors to LNI have been identified. In Chapter 5 we discussed two hypotheses accounting for heteroduplexes that were parental for flanking markers. These hypotheses were patches and clustered chromosome exchanges. Each is a potential contributor to LNI. In addition, we shall have to contend with a third recombinational process, that of heteroduplex correction. It will be our rather awkward conclusion that, to varying degrees in different creatures, all three of these potential contributors play a role in LNI. Sorting out their contributions in any one creature is a miserable job, not yet fully accomplished. In Chapter 5 we presented the argument for patches. Let us now look at the evidence for clustering and heteroduplex correction.

EXCHANGE CLUSTERS

In Chapter 5 we examined branched T4 DNA molecules isolated from infected cells in which DNA replication was blocked. We concluded that these molecules are likely to be recombinational intermediates since a variety of conditions that influence rates of recombination (Chapter 9), as measured by conventional crosses, similarly influence the frequency of branched molecules. Analysis of the distribution of branches indicates nonrandomness. Pairs of branches within 3,000 base pairs of each other were observed in excess of their random expectation.[2] Since branches may be resolvable as either splices or patches, the clustering of branches implies clustering of splatches.

In Chapter 10 we shall speculate on the origin of these clusters in replication-blocked crosses of T4. Clustering in ordinary, fully replicating crosses is deducible from the description of T4 growth and recombination presented in Chapter 4. There we stated three features that, together, predict clusters of splatches: (1) recombination is high at the ends of molecules; (2) the map location of molecule ends vary; and (3) early replication of a chromosome leads to a clone of molecules with ends in about the same places (see Chapter 10). Two (or more) members of such a clone can recombine at one of their active ends with a third molecule, like this:

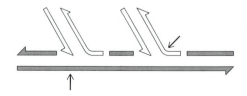

Cuts at the indicated places on this intermediate would lead to a closely paired splice and patch:

which, on one of its chains, has a close triple chain exchange.

MISMATCH CORRECTION

The possibility that heteroduplexes might be "corrected" to homoduplexes[3] is suggested by reactions that repair radiation- or chemical-induced damages in duplex DNA.[4] In some of these repair reactions, enzymes that recognize the damage break an internucleotide link on the damaged chain. Enzymatic hydrolysis of the "nicked" chain then proceeds, removing the damaged bit plus nucleotides for some distance on one or both sides of it. The resulting gap is filled in by enzymatically catalyzed nucleotide polymerization with the intact chain acting as template:

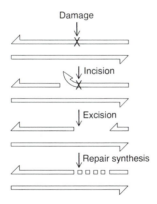

Extrapolating the properties of a damage-repair system to one that corrects mismatches (or nonmatches for the case of deletions or additions) would lead us to expect the following.

1. The reaction that corrects the mismatch should be triggered by the mismatch. Since the reaction will occur only on heteroduplex DNA, it can be said to be caused by the markers used in the cross; i.e., it is a marker effect. More precisely, it is caused by a combination of markers, usually a mutant and its wild-type alternative. Neither of the two homoduplexes will cause the reaction since neither involves a Watson-Crick base mismatch.

2. If the excised stretch of DNA (the excision tract) ends between two marked sites, it may recombine markers present in that splatch, as for example:

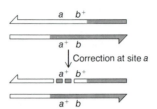

(The broken line indicates DNA synthesized as the repair step in the correction process.) Therefore, mismatch or heteroduplex correction is a marker-stimulated recombination reaction.

3. Note that the ends of an excision tract act as potential points of recombination when they fall within a splatch, and thus they constitute a source of LNI.

4. If sites *a* and *b* in the above diagram are far apart compared to the length of the excision tracts, they will be corrected independently. The recombination frequency due to correction will then be a function only of the two separate repair reactions and will be independent of the distance between the sites.

5. If sites *a* and *b* are about as far apart as the average length of excision tracts, or closer, their corrections will no longer be independent. In the following diagram, correction of the mismatch a/a^+ simultaneously corrects the mismatch at site *b* (co-correction):

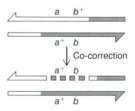

Co-correction gives both chains the same parental genotype. Thus, marker proximity reduces the rate of recombination in mismatch correction reactions, and recombination due to correction becomes distance-dependent.

6. It is almost inconceivable that enzymes could "know" which of the two base sequences in a heteroduplex is the "good" one. Therefore, we may

expect correction in the direction of either the mutant or the wild-type homoduplex. This contention is deliberately ambiguous. It can be interpreted to mean (a) that heteroduplexes at some sites correct preferentially to the mutant while others correct to the wild type, and that we have no basis for prediction, or (b) that at individual sites correction may occur in either direction. Consider, for example, a site in which the wild type has an A:T base pair while the mutant sequence contains G:C. The heteroduplexes resulting from splatch formation will be of two sorts—A:C and G:T. Correction may occur by any or all of four different reactions: A or C may be excised from the A:C mismatch, or G or T may be excised from the G:T mismatch. Two of these reactions (C and G excision) result in correction to the wild type, while the other two yield mutant homoduplexes. Again, we have no apparent basis for prediction; we must not be surprised if correction at such a site occurs in both directions (to wild or to mutant); at the same time, we must not be surprised if the rates of correction in the two directions should turn out to be unequal. For some classes of mutation other than base substitution, our anticipations can be a bit better defined. For instance, heteroduplexes involving deletions or additions might well be corrected in one direction only; it is difficult to imagine that a structure like this

could be corrected with equal probability in either direction.

The occurrence of heteroduplex correction has been established through studies in both phage and fungi. We shall present those for λ here and those for fungi in the next chapter.

Lambda (and some other phages) are better than fungi for the demonstration of correction; in λ, heteroduplexes can be constructed in vitro under well-defined conditions. These heteroduplexes can be introduced into bacteria specially treated to take up DNA molecules from their surroundings. These transfected cells produce phage as if they had been infected by the normal route. Examination of the progeny particles can reveal the consequences of correction. One conveniently monitored consequence, the appearance of recombinants in the progeny, occurs when the transfection is made with doubly marked heteroduplexes in the absence of the ordinary recombination systems. The details of this demonstration are instructive.[5]

The markers used are single-base substitutions that result in conditionally lethal mutations called amber (*am*). These mutations, which render the

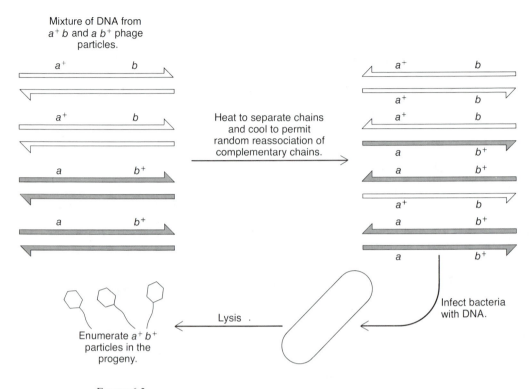

FIGURE 6-2
An experiment to demonstrate mismatch correction in phage λ. (J. Wildenberg and
M. S. Meselson [1975] *PNAS* **72**:2202.)

phage incapable of plaque formation on a restrictive host, are available in
each of the λ genes. Doubly marked heteroduplexes are synthesized by
separating (with heat or alkali) the complementary chains of each chromosome in a mixture of two kinds (Figure 6-2). The resulting mixture of single
chains is then cooled to permit reformation of Watson-Crick duplexes
(see Chapter 9). Half of the reformed duplexes will be heteroduplexes. Note
that there are two kinds of heteroduplexes; one kind is *a* on the Watson
chain and *b* on the Crick while the other is vice versa. The DNA used is
missing its *int* and *red* genes. The reformed duplexes are introduced into
bacteria genetically defective in the Rec system (*recA*). These transfected
cells produced progeny that are examined for the frequency of a^+b^+ (wild
type) recombinants. This frequency is the number of progeny particles that
make plaques on the restrictive host divided by the number that make
plaques on the permissive host, which supports plaque formation by both
am and wild-type particles alike.

Unlike the case with an ordinary phage cross, however, the frequency of recombinants is essentially unrelated to the physical distance between the markers. The following analysis of the data indicates that the frequency of wild-type recombinants is about equal to the sum of the probabilities of correction to wild type at each of the two sites involved in the heteroduplex.

If correction probabilities are small (less than 20 percent or so) and correction at the two sites occurs independently, then the frequency will be $\frac{1}{4}(p + q)$, where p is the probability of correction to wild of the mismatch a/a^+ and q is that for b/b^+. The factor $\frac{1}{4}$ is composed of two factors of $\frac{1}{2}$; half the duplexes are heteroduplexes and half the progeny from a heteroduplex corrected to wild type are wild type. If each of n mutants is used in every possible pairwise combination, the number of frequencies of wild types measured is $n(n - 1)/2$. These measurements provide an equal number of simultaneous equations involving n unknown probabilities. The values that generate the set of frequencies in best agreement with the data can be calculated. The adequacy of the assumption of additivity of correction probabilities is tested by the goodness of the fit. Data from such a test are shown in Figure 6-3, where the fit is seen to be satisfactory. It is probably significant that the four crosses in Figure 6-3 that yielded the fewest wild-type recombinants compared to the predicted values are the ones that are physically closest on the chromosome. Evidently, these sites were co-corrected due to excision tracts beginning at one site and extending across the other. On that assumption, the length of excision tracts in λ is roughly 10^3 nucleotides long. These studies support all the anticipated aspects of correction except for the possibility that the direction of correction (mutant to wild type or vice versa) may be unequal in a given heteroduplex; however, they were not directed to that question. Some of the studies we shall encounter later support that aspect of correction as well.

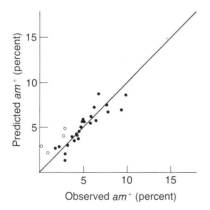

Predicted am^+ (percent)

Observed am^+ (percent)

FIGURE 6-3

Frequencies of wild-type λ from transfecting heteroduplexes made by annealing ab^+ chains with a^+b chains, where a and b are different amber mutations. The predicted am^+ frequency is the sum of the correction frequencies of the single sites to wild type as estimated from all possible two-factor heteroduplex transfections. The four open circles represent the four most closely spaced pairs of mismatches. (J. Wildenberg and M. S. Meselson [1975] *PNAS* **72**:2202.)

Patches and splices share a common feature. Each may be subject to marker effects if the markers in use influence heteroduplex formation. However, for very close markers that have no influence on heteroduplex formation, recombination by either patches or splices, whether they be clustered or not, will depend only on (map) distance, and recombination frequencies will be rectifiable to yield a map with additive distances. In short, the recombination rules of Chapter 1 will apply. On the other hand, heteroduplex correction is itself a marker effect. Therefore, recombination resulting from correction may break the rules of Chapter 1. We need a strategy for dealing with these violations. Let us try the following. Any example of recombination of very close markers will first be examined from the point of view of purely distance-dependent reactions. If this proves inadequate, and it sometimes will, we shall call upon heteroduplex correction or other marker effects as needed. We shall start with T4, since almost all the data on this creature can be explained with purely distance-dependent reactions.

LOCALIZED NEGATIVE INTERFERENCE IN T4

In T4, recombination between very close markers has been studied primarily in the *rII* region.[6] Before we look at the data, we should take a moment to consider the suitability of T4 and the *rII* region for such studies.

Short-distance recombination studies in T4 are potentially subject to confusion due to the properties of terminal-redundancy heterozygotes. Consider the two-factor cross $a_1 a_2^+ \times a_1^+ a_2$. About 1 percent of the progeny will be terminal-redundancy heterozygotes of this sort:

This particle has a parental arrangement of markers at each end of the chromosome, and we would rather not score it as a recombinant. But will we? The extraordinary rate of head-and-tail recombination of T4 chromosomes (Chapter 4) guarantees that many of the heterozygotes like the one above will generate recombinant progeny chromosomes in their first cycle of infection on the assay plate. Therefore, whether or not we will score the particle as a recombinant depends very much on both our methods and on the physiology of the *rII* gene expression. For instance, the usual (easy) way to enumerate recombinants between very close markers is to expose a sample of the progeny population to a condition that is restrictive for the two parental types as well as the double-mutant recombinant. The number of recombinants in the sample is then taken as twice the number of progeny that grow (make a colony, plaque, or what have you). With that method, the wild-type frequencies in crosses involving *rII* markers are certainly influenced by terminal-redundancy heterozygotes, if the markers used are in separate cistrons. When a_1 is in the *A* cistron and a_2 is in the adjacent, *B*, cistron, the heterozygote above is capable of going one full productive cycle of infection on the restrictive host by virtue of complementation.[7] In most of these nascent plaques, $a_1{}^+a_2{}^+$ recombinants will arise due to the head-to-tail recombination, and a fully respectable plaque will result. Intra-cistronic crosses are apparently not subject to this complication. It is a feature of *rII* that the gene products must be informed by the infecting chromosome if they are to be effective; recombination (as well as replication) apparently happens too late to do the job.

The discussion above implies a weakness in many recombination studies with very close markers, in fungi as well as in phage. It is usually the case that the double-mutant recombinant can be distinguished from parental types only by laborious *backcrosses* to each of the parental types. Many workers have therefore taken the easy route of assuming that the two complementary recombinant types occur with equal frequency (Rule 1, Chapter 1) and measure only the wild-type frequency. Barring technical problems, that is acceptable wherever recombination is exclusively distance dependent. Nevertheless, the frequencies of both recombinant classes have been measured in some studies of T4.[8] Whatever technical problems plague the enumeration of one class, enumeration of the other is not apt to be subject to the same difficulties. Thus, it was comforting to find that in T4 the two recombinant classes were about, though not exactly, the same.[9] (Later we shall discuss an interesting exception to the approximate equivalence of the two recombinant classes in T4).

Another caveat relates to heteroduplex heterozygotes. How will such a chromosome be scored and how should we like it to be scored? Consider the usual method of enumerating recombinants by plating a sample of the progeny under restrictive conditions. With very close markers a significant fraction of the chromosomes may be (and in T4 are) recombinant on one chain but not the other, due to splatches in which one end falls between the markers,[10] as for example:

Will such a chromosome be viable under restrictive conditions? That depends on the physiology of the gene used in the studies. In some cases the restrictive conditions are "leaky" enough to permit chromosome replication and the consequent production of a fully wild-type chromosome. In general, a plaque or colony will then result. In the case of rII in T4, half of the chromosomes of the type diagrammed will make a plaque on the restrictive host. Those that are $a_1{}^+a_2{}^+$ on the transcribed chain will plate as wild type; those that are mutant on the transcribed chain will not. This is serendipitous. Since the molecules that are heteroduplex at one site are recombinant on only one of their two chains, it seems fair that exactly half of them be scored among recombinants.

Those readers who have not been intimidated by the above discussion may now want to look at results of three-factor crosses involving very close markers, all in the rII region. In Figure 6-1 the values of "s" were obtained as follows: Each point involves a set of three rII markers; call them a_1, a_2, and a_3. r_{12} is twice the wild-type frequency from the cross $a_1a_2{}^+ \times a_1{}^+a_2$; r_{23} is from the cross $a_2a_3{}^+ \times a_2{}^+a_3$. The frequency of doubles is twice the wild-type frequency from the cross $a_1a_2{}^+a_3 \times a_1{}^+a_2a_3{}^+$. "$s$" is the ratio $r_{\text{doubles}}/r_{12}r_{23}$. Note that "$s$" rises as r decreases, reaching values about 10 times that which can be accounted for by the mating theory.

The results of four-factor crosses argue that the number of chain exchanges on at least one chain can excede two. The triple recombinant occurs in excess by almost the same factor that the double recombinant does in the three-factor crosses.

In principle (see Chapter 1), the coefficient of coincidence can be measured from the results of the three two-factor crosses. This is not easy, however, for very close markers (or even ordinarily close markers). An estimate of the coefficient depends on measuring a deviation from additivity. In the equation

$$r_{13} = r_{12} + r_{23} - 2 \text{"}s\text{"} r_{12}r_{23}$$

the product $2\text{"}s\text{"}r_{12}r_{23}$ must be appreciable compared to the sum $r_{12} + r_{23}$ in order for a deviation to be observed. When r_{12} and r_{23} are small, however, their product is smaller yet, and only a very large "s" will make the last term appreciable. Nevertheless, in T4, two-factor crosses reveal a departure from additivity, indicating a large "s" for very close markers[11] (Figure 6-4). Thus, results of T4 crosses are not inconsistent with distance-dependent recombination mechanisms, so we succumb to the temptation to write a mapping function that incorporates LNI.[12]

Adapting Equation 2-6 to acknowledge clustering of splatches requires that we replace d and $1 - d$ in the exponents with an R that describes clusters. One simple cluster model supposes that splatches cluster because pairing occurs effectively only over short regions.[14] If most of these clustered events are splices, then chain exchanges on each individual chain may be

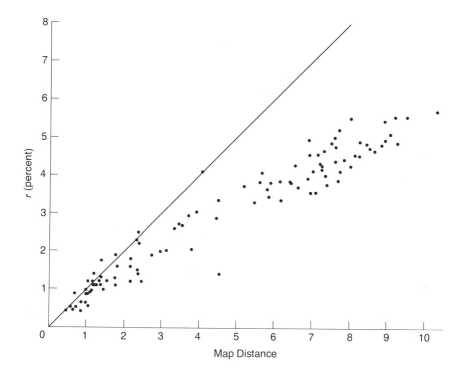

Figure 6-4
Localized negative interference in two-factor intracistronic crosses in the *rII* region of T4.
Deviation of the recombinant frequencies (*r*) from the sums of the frequencies for the
inclusive intervals (map distance) indicates an excess of multiple exchanges over that
expected if exchanges were independent. (K. M. Fisher and H. Bernstein [1965] *G* **52**:1127.)

Poisson-distributed among and within the effective pairing regions. Since
we are ignorant of the true ratio of splices to patches, let us write our function
with the algebraically simple assumption that patches do not occur. It
cannot make much difference to the function, and we do preserve the basic
feature of clustered, distance-dependent recombination events. Also, since it
looks as though heteroduplexes will be fairly enumerated (at least when
using markers in the same *rII* cistron), we shall not distinguish them from
full recombinants. As in our earlier models, *R* depends on an odd number
of exchanges between the markers. These considerations yield the expressions:

$$R_d = \frac{1}{2}(d - k)(1 - e^{-2y}) + k\left(1 - \frac{1 - e^{-2y}}{2y}\right), \text{ when } d > k$$

and $\qquad$ (6-1)

$$R_d = \frac{1}{2}(k - d)(1 - e^{-2dy/k}) + d\left(1 - \frac{1 - e^{-2dy/k}}{2dy/k}\right), \text{ when } d < k$$

where k is the length of the pairing segment (as a fraction of T4 total length) and y is the average number of exchanges per segment. The parameters $m_1 = 9$, $m_2 = 20$, $k = 1.4 \times 10^{-2}$, and $y = 3$ were selected (by trial and error on a computer) to give optimal additivity of the derived d values. The function is graphed at these parameter values in Figure 6-5. The adequacy of the function (Figure 6-6) should not be taken as strong support for the model on which it was based, since other models with similar features yield almost equally good functions. The importance of the function is primarily its usefulness in extending the range over which we can construct rectified maps for T4 (Figure 6-7). Its secondary usefulness is in estimating the total

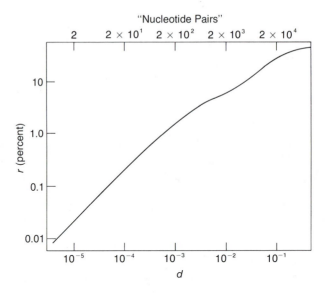

FIGURE 6-5
Combined Equations 2-6 and 6-1 evaluated at the "best" values for its parameters. d is distance on a map of unit length; r is frequency of recombinants; "Nucleotide Pairs" is the number of nucleotide pairs corresponding to d under the assumptions that the T4 chromosome is 2×10^5 nucleotide pairs long and that there is strict proportionality between map distances and physical distances. (F. W. Stahl, R. S. Edgar, and J. Steinberg [1964] *G* **50**:539.)

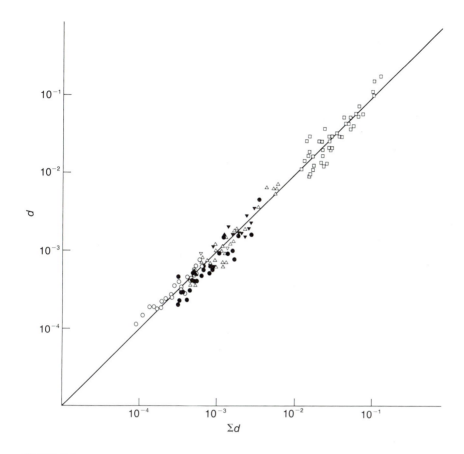

FIGURE 6-6
Degree of additivity of map distances obtained by applying the mapping function in
Figure 6-5 to T4 recombinant frequencies. (F. W. Stahl, R. S. Edgar, and J. Steinberg
[1964] G **50**: 539.)

number of exchanges (43.5 per lineage) and warning us that these are quite
nonrandomly dispersed both among lineages and along chromosomes.
These physical statements based on genetic data accord in a qualitative way
with the clustering observed by electron microscopy. Now let us check two
aspects of our quantitation for consistency.

In the mapping function (Figure 6-5) the length of an effectively paired
segment was estimated to be 1.4 percent of the chromosome length (= 2,800
base pairs). Though the estimate cannot be considered critical, we may still
ask whether it is more or less in accord with another estimate of the length

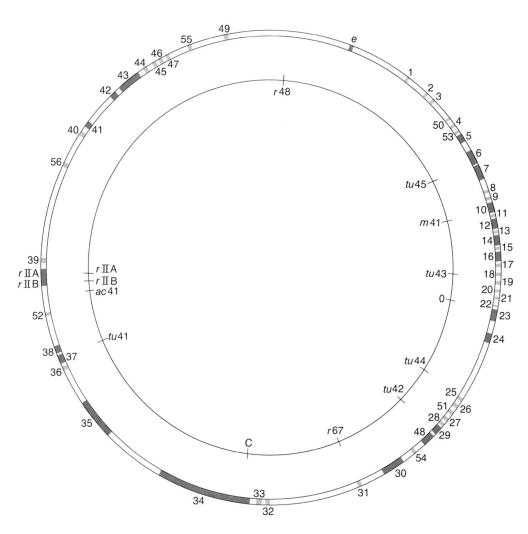

FIGURE 6-7

A rectified map of T4 based on the *am*, *ts*, and *rII* markers used in Figure 6-6. Values of *r* were converted to *d* using the mapping function in Figure 6-5. The elemental *d* values were then adjusted to give $\Sigma d = 1$, then converted to degrees of arc for map construction. Black areas indicate minimal lengths for genes; gray areas indicate locations of genes for which intragenic mapping data were not available. The inner circle shows primarily the locations of plaque-morphology markers like those in Figure 2-1. (F. W. Stahl, R. S. Edgar, and J. Steinberg [1964] *G* **50**:539.)

of paired segments in T4. In Chapter 2 we discussed a pair of crosses that emasculated the mating theory for T4. These same sorts of crosses provide one definition of the size of the pairing region. The experiment asks how long the deletion in the third parent must be before it ceases to decoy the other two. The answer is ". . . larger than the deletion *r196* (100–200 base pairs long) and probably smaller than the deletion *r1272* (about 3,000–5,000 base pairs long)."[15] The joyous fact of consistency is dimmed by the imprecision of the methods. We should note that the range 100–5,000 includes estimated splatch lengths as well, and thus we must wonder whether that is the length of a "mated segment."

In Chapter 2 we examined a T4 mapping function derived in ignorance of LNI. The recombinant frequencies observed for markers across the circle from each other led to an estimate of average "matings" per lineage of $\bar{m} = 6$ (see Problem 2-10). In fitting our function that acknowledges LNI, the computer selected $(9 + 20)/2 = 14.5$ for $\bar{m}$. The approximately twofold discrepancy in these estimates is consistent with the differing intent of the two functions. The first function counts only "matings" in which the number of exchanges in a cluster is odd, while the second function counts odds, evens, and zeros. With an estimated three exchanges per cluster assumed to be Poisson-distributed, the fraction of matings detected with markers that are not very close is $\frac{1}{2}(1 - e^{-6}) \simeq \frac{1}{2}$. We conclude that distance-dependent recombination events, highly clustered, account well enough for T4 recombination with distant, close, and very close markers. Below we shall discuss recombination involving "extremely" close markers, which is not accounted for by distance-dependent mechanisms.

MAP EXPANSION IN T4

In studies involving *extremely* close markers (in which the distances are about codon length), a mapping anomaly sometimes arises in T4.[16] Consider the three possible two-factor crosses involving markers a_1, a_2, and a_3. When these markers are extremely close, r_{13} (which is twice the wild-type frequency in the cross $a_1 a_3^+ \times a_1^+ a_3$) is found to be greater than $r_{12} + r_{23}$. As we discussed in Problem 1-5, such a result suggests that mutation a_2 is a deletion. In this case, however, the coding properties of the mutant strains demonstrate that all three mutations are single-base substitutions. It follows that the markers employed in the crosses are themselves influencing the events leading to recombination; i.e., the genetic distances are distorted by the markers. The map expansion ($r_{13} > r_{12} + r_{23}$) results from finding fewer recombinants for extremely short intervals than are expected from an extrapolation of the function in Figure 6-5 to distances of a few nucleotide pairs, with the shortest intervals suffering more than the longer one.

Can mismatch correction be responsible for this marker effect? Maybe, but it seems more likely that chain exchanges have difficulty occurring between extremely close markers. (In fungi, map expansion occurs for very close as well as extremely close markers, as we shall discuss later.) Is there any need to invoke heteroduplex correction in T4? Maybe. When deletions have been used as markers, two kinds of marker effects peculiar to them have appeared.

For deletions larger than 10–20 base pairs, FUDR (Chapter 5) does not increase the heterozygote frequency.[17] Apparently, such mutations can exist as heterozygotes only in the form of terminal redundancies. (Smaller deletions show FUDR-stimulation of heterozygote formation to an intermediate degree, while deletions of one or two base pairs are found in heteroduplexes at the same frequency as single base-pair substitution mutations.) These observations have implications for the understanding of very close marker recombination in T4. Either large deletions do not enter into heteroduplexes, or else such heteroduplexes are invariably corrected. Two observations support the latter view. (1) In mixed infections in FUDR between nondeleted and deleted strains, the yield of particles favors the deletion by about 3:2.[18] Single base-pair mutations, either substitutions or deletions, enjoy no such advantage. This result can be taken to mean that deletions do enter heteroduplexes and that these heteroduplexes are always corrected in favor of the deletion. (2) LNI has been measured in three-factor crosses both when the central marker is a point mutation and when it is a deletion, as in the following:[19]

$$
\begin{array}{ccc}
a_1 & a_2{}^+ & a_3 \\
\hline
 & \times & \quad \text{Point mutation} \\
\hline
a_1{}^+ & a_2 & a_3{}^+
\end{array}
$$

$$
\begin{array}{ccc}
a_1 & a_2{}^+ & a_3 \\
\hline
 & \times & \quad \text{Deletion} \\
\hline
a_1{}^+ & a_2{}^{del} & a_3{}^+
\end{array}
$$

When the frequencies of double recombinants are taken as twice the $a_1{}^+ a_2{}^+ a_3{}^+$ frequency, then the value of "s" is much less in the second cross than in the first. However, when the frequencies of doubles are taken as twice the $a_1 a_2 a_3$ frequencies, "s" is about the same in the two crosses. The

following is an attractive interpretation. Most very close double recombinants arise via patches, like this

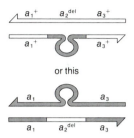

or this

In the upper diagram correction of the heteroduplex in favor of a_{del} restores
a parental genotype, whereas in the lower diagram it generates a double
recombinant. The case for correction of deletion heteroduplexes in T4 thus
looks good. At the same time, work with point mutants, i.e., single base-pair
substitutions or frame-shift deletions and additions, generates data that
can be understood without invoking correction.

LOCALIZED NEGATIVE INTERFERENCE IN FUNGI

Several studies in fungi have been designed as follows. Three linked genes
are identified and their order determined. In the ideal case the three genes
are close, but not very close, and rather evenly spaced, as for example

$$
\vdash\!\!-0.05\!-\!\vdash\!\!-0.05\!-\!\dashv \quad R_{AB} \cong R_{BC}
$$
$$
A \qquad\quad B \qquad\quad C
$$
$$
\vdash\!\!-\!\!-0.10\!\!-\!\!-\!\dashv \quad R_{AC} \cong R_{AB} + R_{BC}
$$

A number of individual mutations, separable by recombination, are then
identified in B. Crosses are made between different pairs of B markers in
the presence of "flanking" markers in A and C, as for example

$$
a \qquad b_2{}^+ b_1 \qquad c
$$
$$
\times
$$
$$
a^+ \qquad b_2 b_1{}^+ \qquad c^+
$$

Haploid meiotic products that have enjoyed a postmeiotic mitosis, and are therefore heteroduplex-free, are plated on a medium that selects $b_1{}^+b_2{}^+$ recombinants. They are enumerated and scored for the frequency of the four possible genotypes with regard to A and C. The a priori expectation in these studies is that almost all $b_1{}^+b_2{}^+$ will be a^+c due to the proximity of A and C and the small S values that characterize crosses with close markers in fungi. More precisely, the a priori expectation in these crosses is that either a^+c or ac^+ will be overwhelmingly the predominant type. If the latter were observed, the cross would be rediagrammed after the fact as

$$
\begin{array}{ccc}
a & b_1\,b_2{}^+ & c \\
\hline\hline
\end{array}
$$
$$\times$$
$$
\begin{array}{ccc}
a^+ & b_1{}^+b_2 & c^+ \\
\hline
\end{array}
$$

However, the typical outcome of such a cross is that all four genotypes with respect to A and C occur with appreciable frequencies among the $b_1{}^+b_2{}^+$ recombinants; instead of being positive, interference is strongly negative.

The frequencies of the AC genotypes vary from one study to another both within and among species. My impressions of these variations are as follows: (1) The parentals $(a^+c^+ + ac)$ are generally equal to or more frequent than the recombinants $(ac^+ + a^+c)$. (2) The parentals may not be equal to each other $(a^+c^+ \neq ac)$; the recombinants, if they approximate half of the $b_1{}^+b_2{}^+$, are even more unequal $(ac^+ \neq a^+c)$. (3) The magnitudes of the inequalities in (2) are correlated between different $b_ib_j{}^+ \times b_i{}^+b_j$ crosses and depend, at least to some extent, on the positions of the marked sites within the gene.[20] It is this systematic quality of the data that encourages interpretation on the basis of distance-dependent models. Two classes of models— the fixed pairing segment and fixed starting point models—have been invoked. We shall discuss the older of the two first.

FIXED PAIRING SEGMENTS

The concept of fixed pairing segments (FPS) supposes that effective pairing segments, which we took to be random in location when we constructed a mapping function for T4 (Equation 6-1), occur in predetermined positions in fungi.[21] They may have both of their ends fixed, one end fixed, or their middle fixed with both ends varying. In any single meiosis only a fraction of the segments pair; positive interference prevents the pairing of nearby segments. Ignoring the duplex nature of the chromatids, we can diagram pairing in a fixed segment, in which both ends are fixed, in this way:

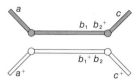

Now we suppose that a variable number of exchanges occurs in the FPS. A $b_1{}^+b_2{}^+$ recombinant will arise if an odd number of exchanges occurs between the B markers. This recombinant will be parental for AC if the *total number* of exchanges in the FPS is even; it will be recombinant for AC if that number is odd.

In the diagram above I have drawn the B markers close to a fixed end of a pairing segment in order to better illustrate the discussion. Since we are confining our attention to $b_1{}^+b_2{}^+$ progeny, each of the paired segments illustrated must have enjoyed an exchange between the B markers. Additional exchanges may occur left of b_1, right of b_2, or both. Were we to observe equality in the frequencies of all four types with respect to the AC markers, we could hypothesize a large number of exchanges within each paired segment. However, the observed inequalities suggest a modest average number of exchanges, not more than one or two. Now let us consider the ways in which a few exchanges will generate the sorts of frequencies observed.

When a parental type for AC is formed, we must suppose an odd number of exchanges outside the b_1b_2 interval. In keeping with the idea of a few exchanges per paired segment, we will consider just a single exchange beyond that which recombined $b_1{}^+$ with $b_2{}^+$. With b_2 close to a fixed end on the right, the exchange outside of b_1-b_2 will usually occur left of b_1. Among parental types for AC, then, ac will outnumber a^+c^+. In cases where parentals for AC are more frequent than recombinants, we must suppose a tendency toward an odd number of exchanges outside b_1b_2. In cases where parentals are about equal to recombinants, we must have zero or an even number of exchanges outside b_1b_2 with a frequency comparable to odd numbers. How will these be disposed in the segment? If there are no exchanges outside b_1b_2, the recombinant a^+c will be formed. If there are two exchanges, then a^+c will result if both are left of b_1; but ac^+ will be formed if one is left of b_1 and the other right of b_2. With b_1b_2 located as drawn, we would find $a^+c > ac^+$ even with a fairly high average number of exchanges per paired segment. As we consider different b_ib_j located ever further from the right end of that segment, we would find diminution of the inequalities in both the AC parentals and AC recombinant types. (Such a correlation is like the one seen in some studies.[22]) With both ends fixed, the inequality in parental types reverses sign as $b_i b_j$ move far to the left, while the recombinant inequality remains always in favor of $a^+ c$.

The predictions for FPS's with one or two variable ends are similar, but not identical, to those for FPS's with two fixed ends. Consider the case for one fixed end (we draw two diagrams to illustrate the variability of the other end):

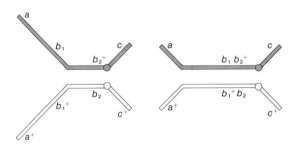

In the first diagram the variable end of the FPS falls between the B markers, whereas in the second both B markers are paired. In this version of the model there is a tendency for $a^+ c^+ > ac$ among $b_1{}^+ b_2{}^+$ simply because the failure of pairing to extend left of b_1 in some instances forecloses the opportunity of an exchange there. As in the version with both ends fixed, the two inequalities $a^+ c^+ \neq ac$ and $ac^+ \neq a^+ c$ vary coordinately as the B marker pair is moved leftward.

FIXED STARTING POINT FOR SPLATCH FORMATION

The second class of distance-dependent models for the segregation of flanking markers among very close recombinants supposes that splatches are initiated at any number of fixed starting points along the chromosome, extending over variable distances in one or the other direction or both.[23] Consider the typical case in which the origin of splatch formation is outside the interval b_1-b_2, i.e., where the dot is in the above diagram. When the splatch spreads left, it will recombine the B markers on those occasions when it stops with its left end between them. When the splatch turns out to be a splice, flanking markers are recombined; when the splatch is a patch, they are not:

Very Close Markers

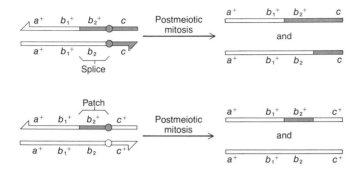

In each diagram the upper product of the postmeiotic mitosis is the one of interest. In the case of the splice, the AC recombinant requiring but a single exchange has been formed. In the case of the patch, one parental type for the AC markers has been formed. How are we to get the other parent, which is generally found in large numbers, and the other recombinant, whose representation is rarely negligible? Let us tackle the ac parental type first. We suppose a second origin of splatch formation to the left of b_1. If it is rather far left, then patches will reach b_1 from the left less often than they reach b_2 from the right, and the ac parental type will be less frequent among $b_1{}^+b_2{}^+$ than the a^+c^+ parental type. As different b_ib_j are used at more leftward positions, the ratios of the two parental types will change, finally favoring the ac one. Fine. Now how does the ac^+ recombinant type arise in the fixed starting point (FSP) model? To answer that, we must look at the results of another cross.

CROSSES WITH AN OUTRIDER MARKER

The following sort of cross has been executed in several fungi with concordant results.[24] In addition to flanking markers at A and C, there is an outrider marker at D. The CD interval is short, but not too short, say $R = 10-30$ percent, a convenient range for evaluating interference.

$$\begin{array}{cccc} a & b_1\,b_2{}^+ & c & d \\ \hline \end{array}$$

$$\times$$

$$\begin{array}{cccc} a^+ & b_1{}^+b_2 & c^+ & d^+ \\ \hline \end{array}$$

Within the class $b_1{}^+b_2{}^+$, those that are recombinant and those that are parental for markers in the flanking genes AC are compared with respect to the recombination frequency for C and D. In the case of parental AC, the recombination frequency for CD is the same as it is in the overall progeny. When AC are recombined, however, the recombination frequency for CD is depressed. Thus, among $b_1{}^+b_2{}^+$ recombined for AC, positive interference is seen in the CD interval, while the event that yields $b_1{}^+b_2{}^+$ without recombining A and C does not interfere with exchanges in the CD interval. This experiment not only tells us where some of the $ab_1{}^+b_2{}^+c^+$ come from but further defines our two distance-dependent models as well.

The result with the outrider marker tells us that the event that forms $b_1{}^+b_2{}^+$ interferes with nearby exchanges only if that event recombines A and C. This result forces a substantial modification of the fixed pairing segment model. We must abandon the assumption that interference in the FPS model enters at the level of pairing between segments. Instead, whenever an odd cluster of exchanges occurs in a segment, interference is imposed on the occurrence of nearby odd clusters, while even clusters impose no such interference. This is a formal possibility that might be more attractive if it could be modeled in molecular terms. For the FSP model, we must say that splices interfere with nearby chromosome exchanges but patches do not. If splices and patches are derived from a common intermediate, as suggested for T4 on page 93, then interference must be imposed at the time of resolution of that intermediate to produce a splice. While failing to explain interference, this conclusion does impose an interesting restriction on FSP models.

Now we can see where at least some of the $ab_1{}^+b_2{}^+c^+$ spores may have come from. In those cases where $b_1{}^+b_2{}^+$ arise by patches, the lack of interference in the AB and BC intervals will allow exchanges unrelated to the patch to occur at a rate that seems "unexpectedly" high, because our experiences with close markers had led us to expect positive interference. These additional exchanges will convert AC parental types to the rare recombinant, like this:

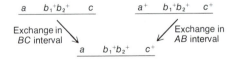

Of course, in the FPS model, production of the ac^+ type via additional chromosome exchange in the flanking intervals must also be considered a source of $ab_1{}^+b_2{}^+c^+$ recombinants. Can the FPS and FSP models be distinguished by random spore analysis of the products of postmeiotic mitosis? Maybe, but certainly not easily. The following two considerations might help.

First, in the FPS model as we have set it forth, there is no special feature to regulate the distribution of exchanges among FPS's. It is thus an economical assumption to let the distribution be a random one—in fact, a Poisson one with a modest mean as we have described. Once we have selected for $b_1{}^+b_2{}^+$, the Poisson-distributed exchanges occurring outside that marked interval will determine whether or not A and C are recombined. In a Poisson distribution the sum of the probabilities for zero plus the even numbers is always greater than the summed probabilities for the odd numbers. Since odd numbers of exchanges in the FPS outside the marked interval are required to restore the parental configuration for A and C, the FPS model in this simple form expects to find AC recombinants more frequent than AC parentals among the $b_1{}^+b_2{}^+$. Although near equality is observed in some cases, e.g., in yeast (see Chapter 8), the general rule seems to be an excess of parental types, and even in yeast there appears to be a slight bias in that direction. If FPS is to be retained as the general basis for LNI in fungi, it appears that we must specify a feature that regulates the number of exchanges in an FPS. If the FPS model is modified to allow only singles and doubles, it becomes almost indistinguishable from the FSP model by random postmeiotic spore analysis.

Second, FSP and FPS with two fixed ends are distinguishable, in principle, if we can locate the fixed points featured in the two models. Among $b_1{}^+b_2{}^+$ recombinants in the diagram on page 109, FPS predicts an excess of ac over a^+c^+, while FSP makes the contrary prediction. This test may become feasible when mutations of the fixed elements become available, allowing their map locations to be determined. In Chapter 9 we shall see that this may be a not too distant eventuality. Efforts to distinguish FPS and FSP models may be doomed by realities that fail to conform to either of the two abstractions. Distinction between the two models is further blurred by the possibility that the chemical basis for pairing is in fact splatch formation. Furthermore, when we examine tetrad data in Chapters 7 and 8, we shall see that the realities of very close marker recombination involve additional phenomena, not apparent in random spore studies.

The FPS and FSP models both postulate special places (ends of pairing segments or starting points for splatches) at fixed points along a chromosome. These fixed places impose the inequality characteristically seen between the frequencies of types parental for flanking markers among very close marker recombinants. Is it possible that this "polarity" is due to features of the locale of the very close markers but instead to some overall polarity of the chromosome, determined perhaps by the centromere or telomere? This possibility was neatly addressed in a cross of a *Neurospora* strain inverted for a marked segment of the chromosome lying between flanking markers.[25] The genotype frequencies indicated that the fixed element that determined polarity had been inverted along with the marked segment and must therefore lie near the very close markers.

The FSP and FPS models are two kinds of distance-dependent models for the analysis of random spore data. Qualitatively, at least, they do the right sorts of things, and all models are likely to incorporate basic features of one or the other, or both. Nevertheless, one feature of random spore data suggests that no purely distance-dependent model will be adequate.

MAP EXPANSION IN FUNGI

In T4 the LNI observed in multifactor crosses was confirmed by two-factor crosses in which $r_{13} < r_{12} + r_{23}$. This has often not been the case for very close markers in fungi. In fact, it is sometimes observed that $R_{13} > R_{12} + R_{23}$, a phenomenon we called map expansion in the case of *extremely* close markers in T4. What are the implications of map expansion for our understanding of very close marker recombination in fungi? First, map expansion tells us that we are plagued by marker effects. Some of them may be due to the inclusion of data from crosses involving deletions (see Problem 1-5), but in other cases that is unlikely. Thus, Rule 3 of Chapter 1, which states that recombination depends on the distance between markers and not on the markers themselves, appears to be violated frequently in crosses of fungi involving very close markers. If that is so, we may be in for serious trouble; genetic analysis of the details of recombination will be perturbed by the very markers we are using to probe those details. Second, the failure to observe LNI in two-factor crosses opens the possibility that some of the observed LNI is in fact caused by the third marker! If such be the case, then all versions of cluster models (with or without pairing segments, fixed or unfixed) and splatch models (with or without fixed starting points) are inadequate for fungi. A leading contender for the marker effect responsible for map expansion in fungi is heteroduplex correction.

As shown in the diagram at the top of page 94, correction within a splatch contributes to LNI. This same process could create map expansion. In the correction reaction, chain excision is presumed to start at a mismatch or nonmatch and to extend for some length in one or both directions. The correction process excises the marker that stimulated it and takes with it nearby markers as well (see the diagram at the bottom of p. 94). If we presume that excision runs in both directions in each act and extends for a fixed length that is shorter than the splatch, then map expansion will occur.[26] Markers

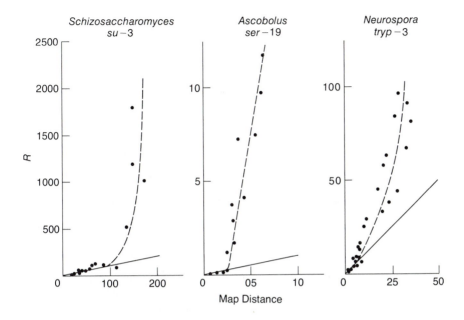

FIGURE 6-8
Examples of map expansion. The map distance is the sum of the recombination frequencies (R) of elementary intervals; R is the observed recombination frequency. The solid lines are the expectations if recombination frequencies are additive. R values are as follows: *su-3*, prototrophs per 10^6 ascospores; *ser-19*, wild-type spores per 10^3 tetrads; *tryp-3*, prototrophs per 10^5 ascospores. (J. R. S. Fincham and R. Holliday [1970] *MGG* **109**:309.)

that are closer together than the excision length will always be co-corrected and hence will be recombined only when the end of a splatch falls between them. Markers further apart than the excision length, but still close enough together to be covered by the same splatch, can recombine in two ways: they recombine when one end of the splatch falls between them and by virtue of correction occurring at the individual marked sites.

If correction accounts for map expansion, then R values for markers that are extremely close, i.e., so close that they recombine only when the end of a splatch falls between them, should be additive. Data reported for fungi appear to bear this out (Figure 6-8). Thus, map expansion in fungi is apparently of different origin than that reported in T4, in which the observed expansion is for intervals one or two bases long.

The assumption of excision tracts of fixed or quasi-fixed length should not be made casually, since it contradicts the simplest model for excision. This model supposes that excision is the result of action of a processive exonuclease (see Chapter 9), an enzyme that removes nucleotides sequentially without letting go of the duplex substrate. Of course, at each step there is a finite probability that the enzyme will release its substrate—its affinity is not infinite. If the enzyme has no memory, i.e., cannot measure how far it has traveled, and if its affinity is independent of nucleotide composition in the neighborhood, then excision tracts will have an exponential length distribution. It is shown in the appendix that such a length distribution does not give map expansion.

Of course, it is possible that the enzyme has a memory, and it is very possible that its affinity varies strongly with varying nucleotide composition. Either of these phenomena could result in more or less fixed-length excision tracts. Nevertheless, the need to make these assumptions leads us to question whether excision tracts can account for map expansion after all.

An alternative possibility is that map expansion is due to man's desire to impose order upon data. This view of map expansion as an artifact says that in crosses with very close markers, marker effects of an as yet undivined nature result in recombination frequencies that bear little relationship to distance. Whether or not map expansion in general is the result of forcing a map on nonmappable data will depend on the distribution of R values that nature presents us with. Problem 6-8 investigates one distribution of R values that frequently leads to the false conclusion of map expansion. The significance of map expansion will be clear when we have measurements of both R values and physical distances for several sets of very close markers.

An unusual marker effect, interpretable in terms of mismatch correction, has been reported for a set of crosses in a *Neurospora* gene (called *am*, unfortunately).[27] The two mutations that manifest the marker effect lie only three base pairs apart in adjacent codons. Both of the mutations arose by transversion, but one was from purine to pyrimidine on the transcribed chain and the other from pyrimidine to purine. (An ingenious interpretation of the marker effect hinges on this bit of intelligence.) Each of the mutants was crossed to markers lying at various distances, all much greater than three base pairs, to one side or the other. The way in which R increased with distance was different for the two mutations. In crosses to markers on the left, one of the mutations gave R values that rose faster with distance than did the other. In crosses to markers on the right, the latter mutation showed R values rising the faster with distance (Figure 6-9). Two other key observations are important to the interpretation: flanking markers were usually in parental configuration among am^+ recombinants, and one parental type was conspicuously more frequent than the other among am^+ recombinants.

The interpretation was made within the framework of a "splatch and mismatch correction" model. The high frequency of parental types for flanking markers indicated that patches were the predominant type of

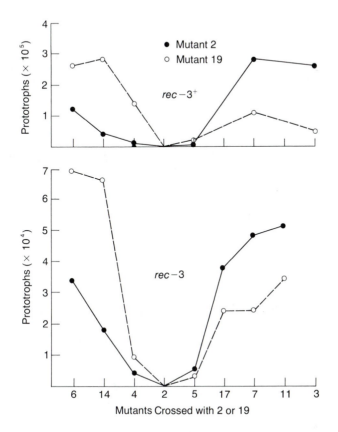

FIGURE 6-9

The wild-type recombinant frequencies shown by mutants *2* and *19* of a *Neurospora* gene, *am*, in crosses with other mutants in the same gene. The mutant sites are indicated along the abscissa in what is thought to be their correct order, but they are spaced at arbitrarily equal intervals. The remarkable marker effect observed in the *rec-3+* strain (*upper graph*) is preserved in the *rec-3* strain (*lower graph*) in which the overall rate of recombination in the gene is about twentyfold higher. The *rec* genes are discussed in Chapter 9. (J. R. S. Fincham [1976] *H* **36**:81.)

splatch. The excess of one parental type over the other indicated that patches made on one chromatid were more correctable to wild-type recombinants than were those on the other. It was then assumed that patches are made on chains of only one chemical polarity, that excision tracts are longer in one chemical direction than the other, and that purine-purine mismatches are corrected to wild type at a different rate than the pyrimidine-pyrimidine mismatches. Convince yourself that this set of assumptions accounts for the phenomena observed in this *Neurospora* gene. Then see if you can find a better, simpler, argument. (The model is set forth in generalized form in Section 6 of the appendix.) The assumption that excision tracts are longer in one chemical direction than the other is supported by experiments on correction of artificial heteroduplexes in λ.[28] The possibility that patches may be formed preferentially on one chain receives support from λ experiments described below. In Chapter 7 we shall examine more direct arguments in support of heteroduplex correction in fungi.

LOCALIZED NEGATIVE INTERFERENCE IN LAMBDA

LNI in λ is illustrated in Figure 2-5. That result tells us little about the mechanism, though one small thought can be squeezed from it. The way in which "s" rises as r gets smaller suggests that "s" might keep going up with ever smaller r. If that happened, we could retain a cluster model only by supposing that clusters have clusters, clusters within clusters have clusters, ad infinitum. Another experiment argues against clusters in λ.[29]

Crosses of the following sort were conducted with both Red and Rec active:

The *bio* mutants are substitutions of bacterial DNA into λ. Their right ends vary in distance from *c*. Progenies of the crosses were plated on a host that allows only the non-*bio*-O^+ recombinant to grow and upon which the *c* and c^+ plaque morphology markers can be distinguished. The ratio c^+/c was found to be proportional to the physical distance of *bio* from *c* (as determined by electron microscopy.) The distances involved extended through the range for which *s* is seen to rise in Figure 2-5. This simple result tells us that when odd numbers of exchanges are selected (in the interval *bio*-*O*) the efficiency with which these recombine *bio* and *c* is independent of distance. If clusters of, say, three exchanges existed, each element would contribute to recombination probability when the distance is small, whereas the efficiency of that cluster would drop threefold at distances larger than the span of the cluster. The data were taken to indicate that among chains with odd numbers of exchanges in the interval studied, more than 99 percent bore single exchanges. Two features of this experiment make its generality somewhat questionable. I am disquieted, without knowing why, by the use of a gross heterology (*bio*) as a marker. Moreover, most *bio* substitutions contain Chi sequences (Chapter 9) for the initiation of Rec-mediated exchange at high rates. The possible activity of Chi in these experiments makes it difficult to compare them with other λ crosses. These results do, however, encourage us to get on with the consideration of other sources of LNI. The crosses described, since they select for odd numbers of exchanges, do not bear on patch formation (or clusters of two splices) in λ. An attempt to demonstrate patches is described below.[30]

The crosses were replication-blocked and density-labeled so that material and informational contributions to recombinants could be assessed simultaneously. Both two- and three-factor crosses with very close markers were conducted and the results were similar. The crosses were as follows:

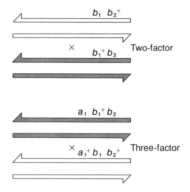

Recombination due to short patches would produce chromosomes that were wild type on one chain (and detectable when that chain was the transcribed one) and almost parental in density. Thus, the two-factor cross would look like this:

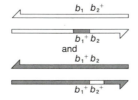

and the three-factor one like this:

$$a_1^+ \quad b_1 \quad b_2^+$$
$$a_1^+ \quad b_1^+ \quad b_2^+$$

The density distributions found for the recombinants are shown in Figure 6-10. Hindsight leads to the conclusion that the experiment may well be incapable of revealing patches; as we discussed in Chapter 4, chromosome exchange is required for encapsidation of unreplicated λ chromosomes, so the heteroduplexes found are likely to be splices whether or not patches can also occur. Thus, the question of patches in λ remains open, but this is a good time to talk about the heteroduplexes that *were* found.

Crosses with very close markers in the DNA replication genes O and P were conducted with additional markers. One gene close and to the left (cI) was found to be heteroduplex in more than half the recombinants. A gene near the right end (R) was heteroduplex in about 30 percent of the recombinants. When both markers were present, some of the recombinants were found to be heteroduplex at *each* of the flanking marked genes.

Examination of the progeny of these heteroduplexes revealed that most of them were segregating the flanking markers in parental combinations, arguing that the multiple heteroduplex was a result of a single long splice.* Thus recombination for the very close markers in genes O and P must have been effected by mismatch correction. Support for that view was obtained from further features of the data.

Of the two parents in the cross, one was marked $O_{29}P_3^+P_{80}$, the other was $O_{29}^+P_3P_{80}^+$, as in Figure 6-10B, and they were differentially density labeled. The cross was replication-blocked and the progeny was displayed according to density. Samples of phage from the heavy and light sides of the recombinant distribution were then adsorbed singly to bacteria nonpermissive for the parental types. Only two kinds of cells produce progeny: those infected by particles that are $O_{29}^+P_3^+P_{80}^+$, at least on the transcribed chain, and those infected by particles that *become* $O_{29}^+P_3^+P_{80}^+$, presumably by correction, after they enter the nonpermissive host. The infected cells were then distributed for single bursts, and the types of particles produced in each burst were determined. Of course, each burst contained many $O_{29}^+P_3^+P_{80}^+$ particles, but in addition some contained mutants.

---

* It appears that heteroduplexes in λ can extend over 20 percent of the chromosome length, a distance of 10^4 nucleotide pairs. In Chapter 10 we will propose that such long splices are atypical and represent a class selected by the replication-blocking conditions of the crosses.

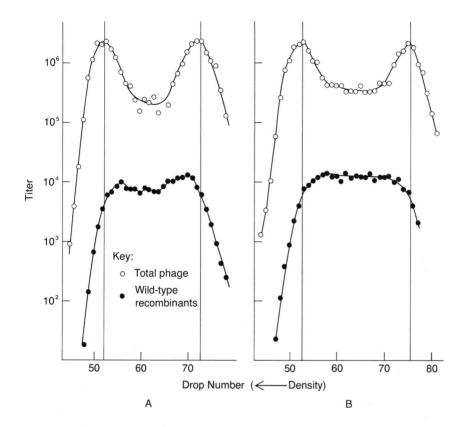

Titer

Drop Number (←——Density)

A   B

FIGURE 6-10
Density distributions for close marker recombinants from heavy-by-light, replication-blocked crosses. (F. W. Stahl and M. M. Stahl [1976] *G* **82**:577.)

In *A* the cross was

$$
\begin{array}{c}
\underline{\hspace{3cm}\;+\quad\;\text{P80}\;\hspace{2cm}}\quad\text{Heavy}\\[2pt]
\times\\[-2pt]
\underline{\hspace{2.5cm}\text{P3}\quad\;+\quad\;\hspace{2cm}}\quad\text{Light}
\end{array}
$$

In *B* the cross was

$$
\begin{array}{c}
\underline{\hspace{2.5cm}+\quad\text{P3}\quad+\hspace{2cm}}\quad\text{Heavy}\\[2pt]
\times\\[-2pt]
\underline{\hspace{1.5cm}029\quad+\quad\text{P80}\hspace{1.5cm}}\quad\text{Light}
\end{array}
$$

FIGURE 6-11
Lambda recombination mediated by Red or Red-plus-Rec yields unreplicated progeny chromosomes to which one parent has made a contribution near the right end of the chain that ends with 5'—OH.

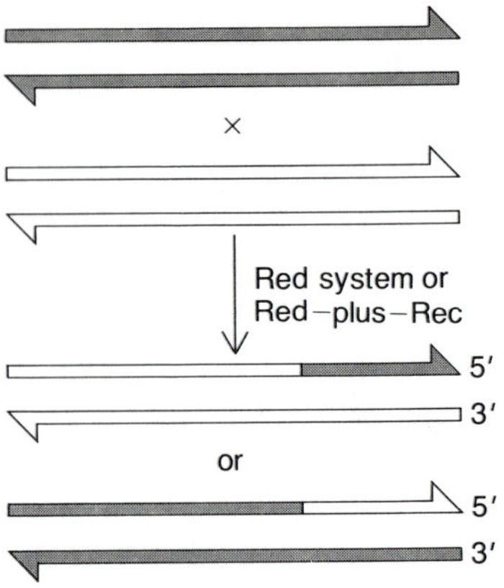

These genetically mixed bursts tell us that the infecting particle retained mutant information on the nontranscribed chain. Remarkably, the recombinants at the light side of the density distribution segregated mutant markers (almost) exclusively from the light parent, while those from the heavy side yielded mutant offspring carrying markers only from the heavy parent. There seems to be only one economical way to interpret this result; the light recombinants have a contribution from the heavy parent on the transcribed chain only, while the heavy recombinants have a contribution from the light parent on the transcribed chain only (Figure 6-11).[31]

Support for the above interpretation is provided by experiments in which heteroduplexes were constructed in vitro.[32] Two kinds were made:

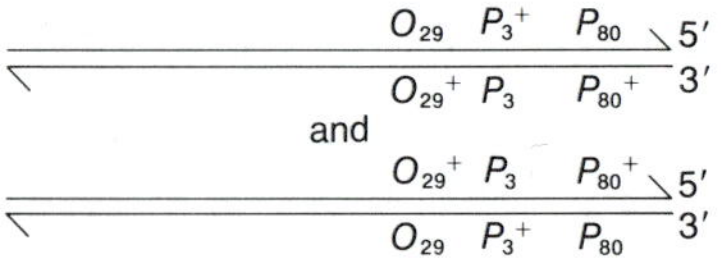

These heteroduplexes transfected nonpermissive cells, and the infected cells were delivered for single bursts. Since both chains of the transfecting chromosomes are mutant in a DNA replication gene, only those cells in which correction occurs will produce progeny. Such cells were found, and the progenies from each were analyzed. The heteroduplexes in the diagram above gave only wild-type progeny in some bursts but wild type plus $O_{29}^{+}P_3P_{80}^{+}$ in others. The heteroduplexes in the lower diagram gave pure wild-type bursts and some bursts with $O_{29}P_3{}^+P_{80}{}^+$. (P_{80} was rarely found.) These progenies were like those found with particles from the density-labeled cross. Thus, they support the interpretation given to those crosses and in addition tell us that many of the particles we called recombinant may have become so only after correction occurred in the host strain that we used to enumerate them. These experiments augment those that used λ to demonstrate heteroduplex correction by showing that the reaction in fact occurs in λ crosses, albeit rather unusual ones.

Several experiments on T4 described earlier in this chapter supported the view that deletion heteroduplexes do arise in T4 and that our failure to find them is due to a high efficiency of correction. For λ, however, deletion heteroduplexes apparently do not form. This conclusion is reached from a variation on the experiment reported in Figure 6-10A. In this experiment the close markers were near the middle of the chromosome, in the bacterial *lac* gene, which had come to reside there. The *lac*$^+$ recombinants arising by recombination between phage strains carrying different *lac* mutations were distributed in a density gradient much like the recombinants in Figure 6-10. When one of the two parents carried either a small deletion or insertion on one side of the marked region, however, the density distribution of recombinants was truncated on that side. When one parent carried a deletion on one side and the other parent a deletion on the other, then the density distribution was truncated on both sides, and few *lac*$^+$ recombinants arose. Those that did arise had a density corresponding to an exchange in the *lac* gene with a splice that was short at best. It appears as though a deletion (when present in one of the two parents) interrupts the formation of splatches that could otherwise have crossed the *lac* gene and been forerunners of *lac*$^+$ recombinants via mismatch correction.[33]

For both T4[34] and λ[35] (and some fungi as well[36]) the results of multifactor crosses with very close markers argue that LNI cannot be entirely accounted for by (nonclustered) patches. If the two ends of singly occurring patches are solely responsible for LNI, then among chromosomes demonstrated to have two close exchanges the probability of a *third* nearby exchange should be normal, showing only the correlation expected from the mating theory. However, the findings are that the correlation is about as high for the third exchange as for the second. This finding also argues against effective pairing segment models in which the number of exchanges in a cluster is not allowed to exceed two. Clusters allowing for greater than two exchanges per cluster and/or mismatch correction appear called for. My favored notions are clusters for T4 and mismatch correction for λ.

PROBLEMS

6-1 Two-factor crosses were performed with markers in the *rII* gene of T4. The following frequencies of wild-type particles were observed in the progenies:

$$a_1 a_2{}^+ \times a_1{}^+ a_2 \qquad 0.010$$
$$a_2 a_3{}^+ \times a_2{}^+ a_3 \qquad 0.0050$$
$$a_1 a_3{}^+ \times a_1{}^+ a_3 \qquad 0.013$$

Calculate "*s*," the coefficient of coincidence for this cross.

6-2 Three marked sites, *a*, *b*, and *c* are very closely linked in that order. The recombination frequencies for the elementary intervals are $R_{ab} = 1.0 \times 10^{-3}$ and $R_{bc} = 2.0 \times 10^{-3}$, respectively. In the three-factor cross $a^+ bc^+ \times ab^+ c$, the $a^+ b^+ c^+$ frequency was measured as 2.0×10^{-5}.

(a) What is the coefficient of coincidence?

(b) What value of R_{ac} would you anticipate from the two-factor cross, $a^+ c \times ac^+$?

(c) Suppose S had been twice as big. What R_{ac} value would you anticipate? Compare this value to that in part *b* and decide whether or not S can be accurately determined from the results of two-factor crosses alone.

6-3 S as a function of R was determined in a wee beastie that manifests LNI in crosses with very close markers and positive interference when markers are somewhat farther apart. Each S value was calculated from a set of three crosses: $m_1 m_2{}^+ \times m_1{}^+ m_2$, $m_2 m_3{}^+ \times m_2{}^+ m_3$, and $m_1 m_2{}^+ m_3 \times m_1{}^+ m_2 m_3{}^+$, called *a*, *b*, and *c*, respectively. In cross *a*, $R_a =$ twice the wild-type frequency; cross *b*, $R_b =$ twice the wild-type frequency; in cross *c*, $R_{\text{doubles}} =$ twice the wild-type frequency. Crosses *a* and *b* were selected to give R values that were about equal. S values were calculated as:

$$S = \frac{R_{\text{doubles}}}{R_a \times R_b}$$

The S values were then plotted against $R_a + R_b = 2R$, and a smooth curve was drawn through the values. The resulting curve is shown in Figure 6-12. Following the procedure described in Chapter 1, bootstrap your way to a graphical mapping function for this creature. (Take as unit distance that distance which gives $R = 0.001$. If you interpret the data and the resulting mapping function in terms of a FPS model, about how long (in map distance units) is a FPS? About how many exchanges occur per FPS?

6-4 DNA from two mutant λ types was melted and mixed, and the mixture annealed. Both types were *red int*, and the annealed DNA was used to infect *recA* bacteria. The wild-type frequencies given below were observed in the progenies:

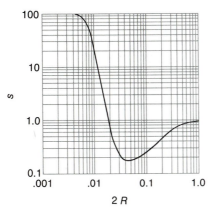

FIGURE 6-12
S versus $2R$ concocted for Problem 6-3.

Source of DNA

$$ab^+ \times a^+b \quad 0.037$$
$$ac^+ \times a^+c \quad 0.030$$
$$bc^+ \times b^+c \quad 0.043$$

(a) From these values calculate the frequencies of single-site correction to wild type for each of the three sites assuming that they are corrected independently and at a low rate.

(b) The assumptions used to make the calculation in part *a* are not tested by the outcome of the calculation. They can be tested, however, if we have a fourth site. Thus, consider a mutant at site *d*, which gave a wild-type frequency of 0.033 when annealed with DNA mutant at site *a*. If the assumptions underlying the analysis of part *a* are valid, what values will you expect in the cases of $bd^+ \times b^+d$ and of $cd^+ \times c^+d$?

(c) DNA of the type $a^+e \times ae^+$ gave a wild-type frequency of 0.050. Calculate the *expected* values for each of the other combinations in the following table:

| | Wild-type frequency | |
DNA	Expected	Observed
$a^+e \times ae^+$	$\times$	0.050
$b^+e \times be^+$	_____	0.021
$c^+e \times ce^+$	_____	0.055
$d^+e \times de^+$	_____	0.058

Explain the discrepancy in the table between one of the expected and observed frequencies.

6-5 In a fungus the following recombination frequencies were measured for three sites linked in the order 1-2-3:

$$R_{12} = 1.0 \times 10^{-4}$$

$$R_{23} = 2.0 \times 10^{-4}$$

$$R_{13} = 4.0 \times 10^{-4}$$

(a) By what name do we refer to this anomalous result?

(b) Calculate the coefficient of coincidence, and think about the value you get.

(c) What mechanisms might have produced the observed anomaly?

6-6 Try to devise an experiment for determining how much of the recombination between close markers is due to splices, patches, and mismatch correction operating on splices and patches.

6-7 I think there is a source of negative interference in phage crosses that has never been identified in the literature. During the period when phages are being matured, DNA replication maintains a "pool" of "vegetative" recombining chromosomes. We may suppose that maturation is at random with respect to genotype. In that case, the ratio of alleles at any locus is subject to random drift. Even in those cells in which the input ratio was close to unity, random drift in the pool may cause the ratio to depart from unity. Since the rate of recombinant production depends on the product of the allele ratios (Equation 2-1), we can expect the rate to fall as cells drift away from initially favorable ratios. I can identify two cases that I would like to see solved.

(a) Consider a phage in which recombinant frequencies are always low (because of small m). For this case I think the contribution of drift to negative interference will enter into the mapping function as a multiplier of r in the same fashion as g, the finite input correction.

(b) This case is more complicated. For a phage like T4, drift of allele ratios at one locus will be more or less independent of drift at remote loci because of the high recombinant frequencies. It seems to me that cell-to-cell heterogeneities resulting from drift will characterize short segments of chromosome. Consider two close intervals. If one of these intervals happens to maintain a favorable (close to 1:1) allele ratio, the second will also (linkage causes it to tend to share a common fate.) Then, the relatively high recombination rate resulting from the lucky lack of drift in one region will be accompanied by a relatively high rate of recombination in regions closely linked to it. I think this will result in a localized negative interference. (Don't look in the back of the book for solutions to this problem; I can't do it.)

6-8 Suppose we have very close sites a-e, for which the various possible two-factor crosses give R values that bear no relationship to distance. In fact, let us go further and suppose that the "R" values for every cross are drawn at random from a bucket of "R" values. Half the "R" values in the bucket, I suppose, fall uniformly in the range 0 to 999 and the other half uniformly in the range 0 to 99. The sum of the two half distributions has 0.55 of its values in the range 0 to 99 and 0.45 in the range 100 to 999. This is the sort of distribution that might obtain if half the crosses were perturbed by an uncontrolled multiplicative variable. You can sample such a distribution pretty well with your telephone book.

Pick a page at random in your phone book. For 10 successive columns, write down the last four digits of the third, fifth, and seventh numbers in each column. These will be the values you assign to the 10 pairwise crosses in the table below. Note that I have already inserted my own set of values (from my telephone book). For each entry in the table I have crossed out the first digit and, when that digit was odd, moved the decimal point one place to the left. Enter your values beside mine and treat them likewise. We call our three sets of data *tel*-3, *tel*-5, and *tel*-7, respectively.

Cross	tel-3 My values	Yours	tel-5 My values	Yours	tel-7 My values	Yours
ab	904.6		978.8		123.0	
ac	900.2		171.2		4855	
ad	399.2		0239		0609	
ae	383.0		4649		131.1	
bc	707.2		541.1		2394	
bd	949.0		6648		343.0	
be	2425		153.3		798.0	
cd	585.1		597.8		309.6	
ce	0079		942.9		0445	
de	139.9		132.4		568.8	

(a) For each of three markers (triads) order the markers by the rule that "*R*" for the outside markers is the largest of the three values. By way of illustration, I got the following sets of values for *tel-3*:

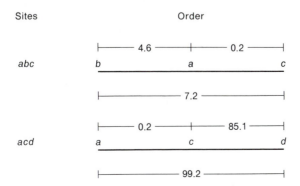

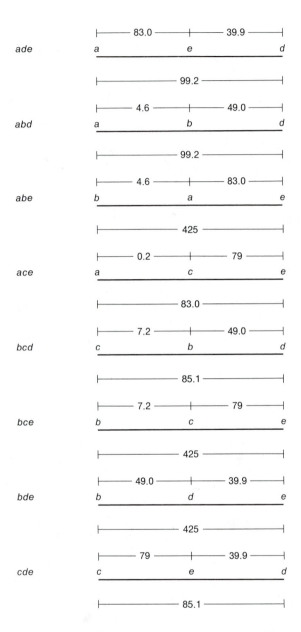

Note that eight of the 10 triads show map expansion. When I examined *tel-5* and *tel-7*, I found seven and eight expansions, respectively (out of 10 each). In aggregate, that gives 23 expansions out of 30, a value significantly different from 15 out of 30 ($p < 0.01$). Do you get a similar result with your *tel* genes?

(b) For each *tel* gene make the best map you can. The map for my *tel-3* looks like this:

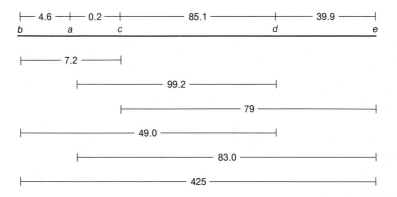

The contradictions are painful, but not much worse than some I have seen for very close marker crosses in fungi. My map for *tel-5* looks like this:

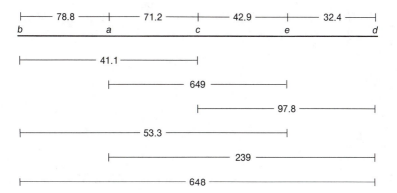

and my map for *tel-7* like this:

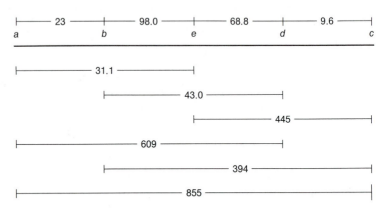

(c) We can test these maps for expansion by plotting "*R*" for each pair of markers versus the sum of the "*R*" values of the elementary intervals between them. For *tel-3* I get the following table of values:

Interval	"R"	$\sum$"R"
bc	7.2	4.8
bd	49	89.9
be	425	129.8
ad	99.2	85.3
ae	83.0	125.2
ce	79	125

In Figure 6-13 I have graphed "R" versus $\sum$"R" for my *tel-3*, *tel-5*, and *tel-7*. Make a graph for your three *tel* genes. Compare your graph and mine with the ones in Figure 6-8. If you find a tendency to map expansion as I did, then you may doubt the significance of that phenomenon.

(Dr. M. Esposito suggested the analysis of the "*tel* genes," though he bears no responsibility for what I have written.)

Now is the best time to study the appendix and to work the problems there.

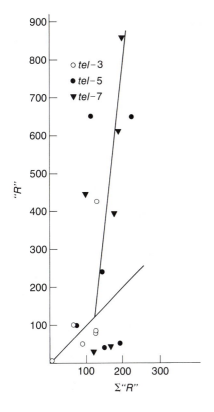

FIGURE 6-13
Artifactual map expansion in the *tel*
genes of Problem 6-8.

NOTES

1. R. H. Pritchard (1960) *GR* **1**:1.

2. T. R. Broker (1973) *JMB* **81**:1.

3. R. Holliday (1964) *GR* **5**:282; H. L. K. Whitehouse and P. J. Hastings (1965) *GR* **6**:27.

4. A. Kornberg (1974) *DNA Synthesis*, San Francisco: W. H. Freeman and Company. G. Pawl, R. Taylor, K. Minton, and E. C. Friedberg (1976) *JMB* **108**:99.

5. J. Wildenberg and M. Meselson (1975) *PNAS* **72**:2202.

6. S. Benzer (1955) *PNAS* **41**:344; M. Chase and A. H. Doermann (1958) *G* **43**:332.

7. F. W. Stahl, H. Modersohn, B. E. Terzaghi, and J. M. Crasemann (1965) *PNAS* **54**:1342; R. Hertel (1965) *ZV* **96**:105.

8. A. H. Doermann and M. Hill (1953) *G* **38**:79; A. H. Doermann and D. H. Parma (1967) *JCP* **70** (Suppl. 1):147.

9. But see K. M. Fisher and H. Bernstein (1970) *MGG* **106**:139.

10. C. Shalitin and F. W. Stahl (1965) *PNAS* **54**:1340.

11. R. S. Edgar, R. P. Feynman, S. Klein, I. Lielausis, and C. M. Steinberg (1962) *G* **47**:179.

12. K. M. Fisher and H. Bernstein (1965) *G* **52**:1127.

13. Stahl, F. W., Edgar, R. S. and Steinberg, J. (1964) *G* **50**, 539.

14. R. H. Pritchard (1960) *GR* **1**:1.

15. Drake, J. D. (1967) *PNAS* **58**, 962.

16. I. Tessman (1965) *G* **51**:63.

17. J. W. Drake (1966) *PNAS* **55**:506.

18. H. Benz and H. Berger (1973) *G* **73**:1.

19. P. Vigier (1966) *CRASP* Ser. D **263**:2010; H. Berger and A. J. Warren (1969) *G* **63**:1; A. H. Doermann and D. H. Parma (1967) *JCP* **70** (Suppl. 1):147.

20. N. E. Murray (1963) *G* **48**:1163.

21. N. E. Murray (1963) *G* **48**:1163; F. Stahl (1961) *JChPh* **58**:1072.

22. N. E. Murray (1963) *G* **48**:1163.

23. R. Holliday (1964) *GR* **5**:282; R. Holliday and H. L. K. Whitehouse (1970) *MGG* **107**:85.

24. D. R. Stadler (1959) *PNAS* **45**:1625; R. K. Mortimer and S. Fogel (1974) in *Mechanisms in Recombination*, R. F. Grell, ed., New York: Plenum, p. 263.

25. N. E. Murray (1969) *G* **61**:67.

26. J. R. S. Fincham and R. Holliday (1970) *MGG* **109**:309.

27. J. R. S. Fincham (1976) *H* **36**:81.

28. R., Wagner, and M. Meselson (1976) *PNAS* **73**:4135.

29. F. R. Blattner, J. D. Borel, T. M. Shinnick, and W. Szybalski (1974) in *Mechanisms in Recombination*, R. F. Grell, ed., New York: Plenum, p. 57.

30. R. L. White and M. S. Fox (1974) *PNAS* **71**:1544; V. E. A. Russo (1973) *MGG* **122**:353.

31. R. L. White and M. S. Fox (1974) *PNAS* **71**:1544; F. W. Stahl and M. M. Stahl in *Mechanisms in Recombination*" R. F. Grell, ed., New York: Plenum, p. 407.

32. R. L. White and M. S. Fox (1974) *PNAS* **71**:1544.

33. E. Sodergren and M. S. Fox, personal communication.

34. M. Chase and A. H. Doermann (1958) *G* **43**:332.

35. P. Amati and M. Meselson (1965) *G* **51**, 369.

36. R. H. Pritchard (1960) *GR* **1**:1.

7

Conversion

In fungi (and presumably other eukaryotes) rare tetrads resulting from
meiosis in a heterozygous diploid contain unequal numbers of the two
alleles. When the eight cells resulting from one round of postmeiotic
mitosis are examined, these exceptional asci may yield six of one kind of
cell and two of the other, or five of one kind and three of the other. This
aberrant segregation implies the gain of copies of information from one
chromosome at the expense of corresponding information on its homolog.
In *Ascobolus*, losses and gains of information are, at least sometimes, the
consequence of mismatch correction of heteroduplexes. In yeast, evidence
for mismatch correction is equivocal, and conversion may have a totally
different basis.

CONVERSION IN ONE-FACTOR CROSSES

Mendel's First Law proclaimed the recovery in equal numbers of both
alleles among the germ cells from a heterozygote. In fungi, with the opportunity to examine the four products from individual meioses, the law is seen
to have more than statistical validity; in general, each allele is represented
in two of the four products in *each* tetrad. As described in Chapter 3, this
reflects the orderly behavior of chromosomes in meiosis. We symbolize that
orderly behavior like this:

For our present purposes, we sacrifice no generality by diagramming the case of no exchange between *B* and its centromere. We will also find it convenient to keep the spores in order. When we do wish to indicate recombinational events, we shall choose the two medial nonsisters as the participants.

Exceptions to this 2:2 regularity are encountered.[1] The exceptions of interest in recombination studies are ones that do not signal a breakdown of the orderly distribution of chromosomes. In these interesting tetrads, normal chromosome distribution is attested to by the 2:2 segregation at essentially all marked loci except the one that is seen to be misbehaving. Localized violations of 2:2 segregation are called (gene) conversion.

Conversion is evidently recombinational. If one marked site is segregating three mutant to one wild type, for instance, whereas other sites on the same chromosome are giving normal 2:2 segregation, it is evident that at least one of the members of the tetrad must be a recombinant. And please note that the recombination implied by conversion is necessarily nonreciprocal; in a tetrad in which Mendel's First Law is violated for only one of two marked sites there is no way that the two complementary recombinant types can be represented in equal numbers. It follows that the two parental types must be present unequally, as well. Recombination by conversion is relatively rare for distant and even for close markers. However, for very close markers (as circularly defined), conversion is a major aspect of recombination; very close marker recombination is frequently nonreciprocal.

By several criteria, conversion is related to aspects of recombination that we described in Chapters 1, 3, 5, and 6.

1. Conversion is frequently accompanied by reciprocal recombination of flanking markers. Thus, conversion and chromosome exchange are correlated and may represent two aspects of a single process.

2. Half or more of all conversions are not accompanied by recombination of flanking markers. Thus, conversion and LNI are correlated and may represent two aspects of a single process.

3. Asci that manifest conversion often show postmeiotic segregation at the converted site, from which we make two deductions: (a) conversion and splatch formation are correlated and may represent two aspects of a single process; and (b) we had better stop talking about conversion as a 3:1 or 1:3 phenomenon and describe it instead as it appears after one postmeiotic mitosis has resolved any heteroduplexes. Conversion asci are then found to be primarily of the following sorts:

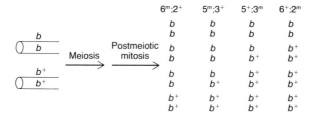

	6^m:2⁺	5^m:3⁺	5⁺:3^m	6⁺:2^m

In the above table the products of postmeiotic mitosis have been kept ordered, so that the top two cells are the mitotic products of the top meiotic product, etc. Note that the 6:2 asci have no meiotic products segregating mitotically and the 5:3 ones have postmeiotic segregation for only one of the four meiotic products.

In addition to the conversion asci and the normal 4:4 ascus, another kind of ascus is sometimes seen. These asci are not conversion asci, by definition, because they contain equal numbers of copies of the mutant and the wild marker. On the other hand, they are not normal either, because two of the four meiotic products give postmeiotic segregation. These aberrant 4:4 asci compare with the normal 4:4 ones like this:

Aberrant	Normal
b	b
b	b
b	b
+	b
b	+
+	+
+	+
+	+

We shall see later that aberrant 4:4 asci are interestingly related to conversion.

Conversion asci giving 7:1 or 8:0 segregation ratios may be encountered. Their low frequency indicates that they are the result of two more or less independent events. As we shall see later, events leading to conversion do not show the interference characteristic of chromatid exchange; consequently, two events in the same locale can be realized. By the same token, postmeiotic segregation in any single ascus is usually confined to one or two of

TABLE 7-1

Data illustrating gene conversion in yeast (*Saccharomyces cereviseae*). Segregation at 30 sites was examined by determining the genotypes of all the spores in 23,135 asci from 20 related hybrids, each heterozygous for 8-13 markers. Data for eight segregating markers are given in detail. The *arg* mutants are probably single base-pair mutations; *ade8-18* may be a deletion. Note that the rate of conversion varies from less than 1 to greater than 8 percent (rates higher than most fungi), that most markers manifest a low frequency of 5:3 segregations relative to 6:2, and that aberrant 4:4 asci are rare. The very rare aberrant 6:2 asci show postmeiotic segregation for two of the four spores. Occasional 8:0 segregations, the result of premeiotic exchanges, have been eliminated from the table.

Segregating marker	No. of segregations	$6^+:2^m$	$6^m:2^+$	$5^+:3^m$	$5^m:3^+$	aberrant 4:4	aberrant 6:2	7:1	abnormal seg. (%)
pet 1	4,924	4	27	0	0	0	0	0	0.63
trp 1	20,826	62	47	12	8	0	0	0	0.63
mat 1	23,135	93	104	0	0	0	0	1	0.85
ura 3	2,315	10	9	2	1	0	0	0	0.95
ade 6	1,589	4	9	3	3	0	0	0	1.20
his 5−2	2,315	14	10	4	1	0	0	0	1.26
tyr 1	8,391	59	44	5	1	0	0	0	1.30
CUP 1	18,016	123	124	6	3	0	0	0	1.42
gal 2	2,416	36	17	0	0	0	0	0	1.78
leu 2−1	3,203	41	25	0	0	0	0	0	2.06
trp 5−48	2,315	23	31	1	0	0	0	0	2.38
met 1	1,589	18	17	18	6	0	0	0	3.72
met 10	892	17	16	1	0	0	0	0	3.82
ura 1	15,014	331	275	11	15	1	0	0	4.23
ilv 3	9.487	230	239	9	6	0	0	0	5.10
lys 1−1	2,315	51	78	3	7	0	0	0	5.76
SUP 6	892	22	33	0	0	0	0	0	6.16
thr 1	21,220	691	594	14	22	0	0	0	6.23
his 4−4	12,533	411	303	39	41	1	0	1	6.36
ade 8−10	1,118	48	30	0	0	0	0	0	7.41
met 13	10,454	459	474	1	· 0	0	0	0	9.07
ade 8−18	15,480	351	239	375	294	10	4	3	9.72
cdc 14	892	48	39	1	1	0	0	0	9.99
ade 7	1,028	47	27	11	8	0	0	0	10.7
his 2	2,481	215	231	2	0	0	0	0	18.2
arg 4−4	1,188	5	13	0	1	0	0	0	1.6
arg 4−3	2,405	28	23	5	3	0	0	0	2.45
arg 4−19	5,352	87	68	2	6	0	0	0	3.04
arg 4−16	14,490	508	302	95	289	4	1	1	8.29
arg 4−17	13,476	485	551	38	20	0	0	0	8.12
Total	221,751	4,521	3,999	657	740	16	5	6	

Source: S. Fogel, R. Mortimer, K. Lusnak, and F. Tavares, *CSHSQB* **43**, in press.

the chromatids. In Chapter 8 we shall further pursue relationships between splatches and chromosome exchanges as well as between conversion and exchange. In this chapter we focus on conversion itself and its relationship to postmeiotic segregation.

The four kinds of conversion asci can be classified in two dimensions: (1) two asci, $6^m:2^+$ and $5^m:3^+$, are conversions to mutant type while the other two are conversions to wild, and (2) the 5:3 asci show postmeiotic segregation while the 6:2 asci do not. In various fungi, as described below, some regularities are apparent in the relative frequencies of conversion asci classified in these two ways.

In yeast, conversions to mutant and to wild type for any marker are about equally frequent (Table 7-1).[2] This "parity" holds for those asci that show postmeiotic segregation and those which do not. Thus, in yeast $5^m:3^+ = 5^+:3^m$ and $6^m:2^+ = 6^+:2^m$. Significantly, parity appears to hold even for deletions as big or bigger than an entire gene.[3] A second generalization apparently valid for yeast is that postmeiotic segregation is rather infrequent. For most markers, $6:2 \gg 5:3$, although in a few cases almost half the conversion asci are in the latter class. The reported behavior of yeast contrasts in several respects with that reported for *Ascobolus*.

For *Ascobolus*, many markers violate parity, with some markers giving conversion exclusively in one direction.[4] Furthermore, some markers manifest postmeiotic segregation in many conversion asci whereas others show it rarely if ever.[5] As we shall discuss later, it appears as though aberrant 4:4 asci are a significant fraction of the abnormal asci in *Ascobolus*, whereas they are a negligible fraction in yeast. It is not obvious that a single molecular model will be able to account for the contrasting behaviors of these two fungi, though in Chapter 11 we shall look at one that tries.

CO-CONVERSION

In both yeast and *Ascobolus* the conversion process shows some respect for distance. Markers that are very close together are more often converted together than are those that are further apart. In the cross $b_1 b_2{}^+ \times b_1{}^+ b_2$ conversion at site *1* to wild type is accompanied in a distance-dependent manner by conversion at site *2* to mutant. The preservation of linkage in co-conversion tells us that conversion involves either a segment of DNA or a finite set of very close points. These possibilities can be distinguished by crosses with three markers. These crosses performed in yeast show that co-conversion of two markers is almost invariably accompanied by co-conversion of any marker between them (Figure 7-1). Thus, we may speak of conversion segments.

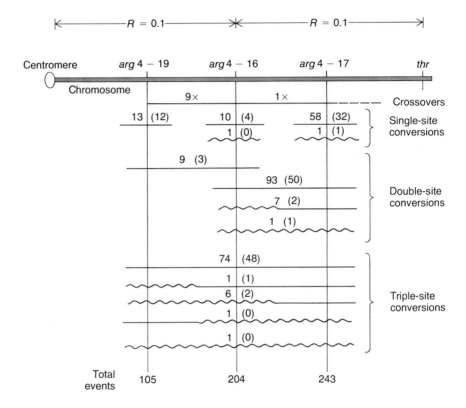

FIGURE 7-1

Pattern of conversions at the *arg4* locus. Three marked sites in the *arg4* gene were flanked by the centromere on the left and a *thr* marker on the right, each about 0.1 units from *arg4*. Among 2,863 tetrads examined, nine were crossovers between *arg4-19* and *arg4-16* while one was a crossover between *arg4-16* and *arg4-17*. All markers segregated normally in those asci. Single-, double-, and triple-site conversions are indicated by horizontal lines. Straight lines symbolize 6:2 asci, while wavy lines stand for 5:3 asci. The number over each line is the number of cases observed, while the number in parentheses reports the cases in which the flanking markers were reciprocally recombined. For instance, the straight line at *arg4-17* is marked 58(32), which means 58 6:2 single-site conversions occurred at *arg4-17*, and 32 of these were crossovers for *arg4-16* and *thr*. The conversion frequencies for the three *arg4* sites illustrate polarity. (Courtesy of F. Tavares and R. Mortimer.)

Models for conversion have been of several sorts. We shall discuss their general features here and then return to particular ones in detail in Chapter 11. To account for the excess of one marker (and its surrounding segment) over that of the alternative, one must postulate DNA synthesis and DNA disposal. Both the gain and the loss of markers has been speculatively ascribed to two contrasting kinds of processes. One class of models says that the synthesis and losses occur here and there in the chromosome complement without regard to any markers that might be about. If the reactions happen to occur in a marked region, we see the consequences of them as gene conversion. The other class of models supposes that the relevant DNA synthesis and degradation are a result of heteroduplex correction; i.e., they are entirely marker stimulated. In Chapter 11 we shall see that some of the most promising models amalgamate these two views.

Demonstrations of heteroduplex correction are a good deal less direct in fungi than in λ. The power of the demonstration in λ resides in the ability to produce heteroduplexes in vitro and to subject them to correction in vivo. Thus, there can be no question that heteroduplexes did exist in λ. In fungi, however, correction has usually been inferred from the failure to find a heteroduplex where some model might expect one. Such a failure, however, has two interpretations: either the heteroduplex *was* corrected or the model erred in supposing a heteroduplex. Though a direct demonstration of correction has been elusive, ingenious experiments in *Ascobolus* leave little doubt about the importance of correction in gene conversion in that fungus.

The demonstration of correction in *Ascobolus* involves the observation that different mutants of a certain gene have very different conversion behavior.[6] The mutants are classifiable according to the frequency with which conversion asci show postmeiotic segregation and according to the degree of departure of conversion asci from parity. Representatives from each class are found throughout a gene, indicating that postmeiotic segregation and disparity are primarily properties of the mutations themselves, not of their location relative to some other chromosomal landmarks (such as fixed starting points for splatches). The validity of the classification scheme is affirmed by the demonstration that certain mutagens induce mutations that fall into only one class while other mutagens yield mutants that are predominantly in other classes. Three classes of mutants have been used in the demonstration of correction. One class showed no postmeiotic segregation and a strong disparity toward conversion to mutant; these mutants gave conversion asci that were exclusively $6^m:2^+$. Another class gave exclusively $6^+:2^m$. With the third class, many conversion asci manifested postmeiotic segregation and the frequency of $5^m:3^+$ was about equal to the $5^+:3^m$. Several properties of these various mutants indicate that those showing no postmeiotic segregation are small deletions and additions of the frameshift sort, while those with postmeiotic segregation and parity are single-base substitutions. From the point of view of the correction hypothesis it is certainly appealing that the grosser violation of the Watson-Crick structure resulting from the nonmatch in an addition or deletion heteroduplex should be corrected so efficiently that postmeiotic segregation is never seen. The strong argument for correction, however, comes from crosses that employ two very closely linked mutations of contrasting conversion behavior.

Two kinds of two-factor crosses were performed.[7] In each case a double mutant was crossed to wild type. In the first cross the double mutant involved one mutation of the exclusively 6:2 variety, while the other mutation was of the 5:3 sort. In the second cross both mutations were of the kind that gave no postmeiotic segregation, but one was $6^m:2^+$ while the other was $6^+:2^m$. In both crosses the proximity of the markers ensured a high frequency of co-conversion. The outcomes of the two crosses were in striking agreement with the simple expectations of heteroduplex correction.

In the first cross the central feature of the data is that the 5:3 site assumed the behavior of the 6:2 site in most asci involving co-conversion. In the second cross the two sites retained their failure to give postmeiotic segregation but each lost its disparity under the influence of co-conversion involving the other. Additional features of the first cross further support the simple view that the 6:2-type site triggers correction with high probability and that the 5:3-type site is simply subjected to the resulting co-correction.

In a small percent of the asci in the first cross, conversion occurred only at the 5:3 site. Presumably, these are the (rare) cases in which a splatch covered only that site. In these cases the behavior of the site was "normal"; i.e., it gave conversion asci with postmeiotic segregation and parity was evident. In a larger fraction of the asci, conversion at the 6:2 site resulted in co-conversion at the 5:3 site, but the latter site still manifested postmeiotic segregation. These are presumably the cases in which a splatch covered both sites, but of the two excision tracts initiated at the 6:2 type site, only one reached across the other site (see below). In these cases, although the latter site retained its postmeiotic segregation, it lost its parity and assumed the disparity of the site influencing it. As mentioned in the paragraph above, in the largest fraction of asci, conversion at the 6:2 site was accompanied by 6:2 co-conversion at the other site.

Is all conversion at these sites due to heteroduplex correction? One feature of the data suggests that it may be. Consider the following simple possibility: splatch formation is reciprocal; i.e., the two interacting chromatids either swap patches of equal length or exchange with the formation of splices of equal length. Among those splatches that cover both sites in the cross, the 6:2 site is corrected to mutant on both chromatids, but the consequences for the 5:3 site can vary. (1) Because of the proximity of the two sites, the most frequent fate of the second site is correction on both chromatids to give a 6:2 ascus. (2) Sometimes the excision tract on only one of the two chromatids reaches the second site; in this case the second site gives a 5:3 ascus. (3) More rarely still, both excision tracts fall short of the second site, and an aberrant 4:4 results. Of course, a 5:3 could result from independent correction at the second site in those cases where neither of the tracts initiated at the first site reached it, but such an event is evidently rare. Thus, the three classes of asci should stand in a particular relationship to each other. If we let p be the probability that an excision tract initiated at the first site reaches the second, then p^2, $2p(1 - p)$, and $(1 - p)^2$, respectively, give the relative frequencies of the 6:2, 5:3, and aberrant 4:4 segregations at the second site. The data obtained fit the relationship nicely when p was taken as 0.93. The result does seem to suggest that splatches are frequently formed on both chromatids and that they may always be so. In Chapter 8 we shall examine a model for conversion and exchange that rests upon the idea that correction acts upon reciprocally produced splatches. However, in Chapter 8 we shall also see that for some loci in some strains of *Ascobolus* splatch formation is probably not always reciprocal. In yeast, there is no evidence that it ever is.

As we mentioned earlier in the chapter, conversion in yeast differs in several respects from that in *Ascobolus*.[8] Conversion in yeast shows few symptoms that suggest correction as the basis, and all evidence suggests that splatches are formed nonreciprocally; only one of the two participating chromatids ever emerges from the conversion process with a splatch. The following evidence supports these contentions. Unlike the case with *Ascobolus*, essentially all mutants in yeast, even gross deletions, show approximate parity. Although this does not argue convincingly against a correction basis for conversion, it does suggest either that correction is, at most, a minor conversion mechanism or that the correction process in yeast is not the sort that I would have envisioned a priori. For most markers, postmeiotic segregation in yeast is rare (Table 7-1). This implies either that correction is efficient or heteroduplexes are infrequently formed. The first possibility is an uncomfortable partner to the parity observation, while the second indicates that we should consider conversion mechanisms other than correction. As with *Acobolus*, the frequency of postmeiotic segregation varies somewhat among markers. However, it is not known whether this variation is independent of position in a way that would clearly signal it as a marker effect. Thus, another kind of marker effect that could implicate correction in gene conversion in yeast is missing. The second piece of evidence is that aberrant 4:4 asci do not occur in yeast except as the coincidence of two independent events. In Chapter 8 we shall see additional data that imply that if conversion usually involves heteroduplex correction, then splatches are not formed reciprocally on the two chromatids whose interaction results in the conversion.

CONVERSION POLARITY

In a number of reports the frequency of conversion at a site is related to the location of that site. In a simple case, the conversion frequency changes monotonically as a function of distance from a point outside the gene. For *Ascobolus*, in which conversion rates vary from mutant to mutant, due most likely to differential correction rates, this dependence on position is not immediately apparent. However, when mutants are first classified according to postmeiotic segregation and disparity, the position dependence becomes apparent within any single class.[9] The nature of the dependence of conversion rate on position is well demonstrated in multifactor crosses of yeast and *Ascobolus*.

In yeast it was demonstrated under conditions in which all conversions at each mutant site were detected. In this case polarity appears to be a rule governing co-conversion. Consider three sites, *3*, *2*, and *1*, linked in that order, subject to co-conversion, and with site *1* showing the highest and site *3* the lowest conversion rate. Conversion at site *3* is generally accompanied by conversion at sites *1* and *2*; conversion at *2* is almost always accompanied by conversion at *1* but not so invariably with conversion at *3*; and conversion at *1* can occur without co-conversion at either *2* or *3*. Thus, it looks as if conversion "started" right of *1* and then extended varying distances leftward, converting everything in its path up to the (variable) point at which it stopped (Figure 7-1).

In *Ascobolus* polarity was first defined with observations made on two-site crosses.[10] Consider the cross $b_1 b_2^+ \times b_1^+ b_2$, in which the mutants b_1 and b_2 as well as the double mutant $b_1 b_2$ produce colorless ascospores. Asci are identified in which conversion at site *1* or site *2* resulted in a colored, wild-type spore. Since co-conversions do not yield wild type, they are overlooked in this analysis. Among the single-site convertants yielding wild type, polarity is manifested as a greater rate of conversion at one site than at the other. This method, which historically defined conversion polarity, is a sensitive, convenient detector. When polarity is present, most of the conversions at one site (say *2*) are accompanied by co-conversion at the other (site *1*), so that most of the wild type that do arise by conversion owe their origin to conversion at site *1*.

For markers within the same gene, it is usual for recombination to be nonreciprocal, i.e., to involve conversion. Thus, a complete description of recombination of very close marker requires that the data be collected in octads, where conversion and postmeiotic segregation can be assessed, and in the next chapter we shall discuss such studies. First, however, let us consider a few hazards.

In some fungi, crosses involving very close markers show strong marker effects, some of which appear to reflect the peculiarities of heteroduplex correction. In these situations recombination frequencies are dubious metrics of distance; in cases where intragenic maps have been based on recombination data, the order of the sites is open to question. Our efforts to understand very close marker recombination would be more soundly based if the order of the markers could be established by independent means. For instance, a map of order and distances could be established by examination of amino acid replacements in the protein products of the various mutants to be used in the recombination studies. Systems suitable for such studies are now becoming available.[11] Meanwhile we take some assurance from claims that determination of the order of mutant sites from recombination frequencies sometimes proceeds without internal contradiction. That is, it seems perturbances in recombination frequencies are not always of such a nature or magnitude as to put in doubt the ordering of the sites.

Somewhat more frightening than the uncertainties introduced by the markers we use is the possibility of cryptic mutations in the gene within which recombination is being studied.[12] Examination of natural populations has revealed that wild-type alleles at any locus are in fact an array of types, all functional, but differing in a base pair or two here and there.[13] In some fungi mating restrictions require that crosses be carried out between descendents of two separately isolated individuals. For any gene there is a real chance that the wild-type states are different in the two clones. Cryptic mutational differences between the two strains can intrude into recombination studies in either of two ways. (1) A mutation that is cryptic in the wild-type gene may alter the expression of markers used in the recombination studies. (2) The cryptic mutation may influence recombination behavior of nearby markers. These potential complications can be ameliorated by deriving stocks of similar genetic background through the technique of backcrossing. Even though we cannot stop this race for a little rain, we ought to proceed under the yellow flag as we try to relate conversion and exchange in Chapter 8.

PROBLEMS

7-1 (a) Does mismatch correction imply a violation of Mendel's First Law (the principle of segregation)?

(b) Can mismatch correction conceivably occur without producing a conversion ascus in an eight-spored ascomycete?

(c) Does the occurrence of conversion asci imply mismatch correction?

7-2 Suppose splatch formation is perfectly reciprocal (which it probably is not, as we shall see later). That is, suppose spliced and patched pairs of chromatids are formed with their points of chain exchange identically placed, like this:

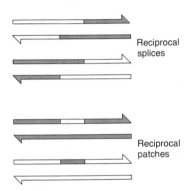

Let mismatch correction be equally likely in each direction in any single heteroduplex and occur with a probability ρ per heteroduplex site.

(a) Consider a one-factor cross. In terms of ρ, what fraction of the tetrads in which a heteroduplex splatch arises will fall into each of the following classes of ascus types: $5^m:3^+$, $5^+:3^m$, $6^m:2^+$, $6^+:2^m$, 4:4 aberrant, and 4:4 normal?

If $\rho = 0.4$, what fraction of aberrant asci will be 4:4 aberrant? If $\rho = 0$, what fraction of aberrant asci will be 4:4 aberrant? If $\rho = 1$?

(b) In yeast, 5:3 asci are seen rather often, but 4:4 aberrants are extremely rare. Relate that fact to the model under consideration in this problem.

(c) Suppose splices and patches are equally frequent. What fraction of aberrant asci will be tetratype for flanking markers? Parental ditype? Recombinant ditype? (For simplicity, let the flanking markers be so close that we can ignore those exchanges between them that occur independently of the ones responsible for the aberrant asci.) Do these fractions apply separately to the 5:3 and 6:2 asci?

7-3 We elaborate on the model of Problem 2. Suppose that R for the interval from the converting site to the left flanking marker is 0.10 and R for the interval from the converting site to the right flanking marker is 0.05. Keep the conventional assumptions of no chromatid interference and no sister-strand exchange. Suppose, in addition, that splices show strong chiasma interference with each other, but that patches are not interfered with by splices or by other patches. Assume as well that the rates of aberrancies for the flanking markers are negligible.

(a) What fraction of the aberrant tetrads are tetratype? Recombinant ditype? Parental ditype?

(b) For 6:2 asci, what fraction of the tetratype tetrads will be recombined for flanking markers for two chromatids of identical genotype for the converting site?

7-4 Suppose all features of the model in Problems 7-2 and 7-3 did describe events properly for some creature, except for the feature that splices and patches are equally frequent. Now suppose that almost exactly half of the 6:2 asci in this creature manifested recombination for the flanking markers in such a way that the chromatids that were recombined were identical in genotype at the converting site. Make the same assumptions regarding R values and interference as we made in Problem 7-3. What is the actual fraction of splatches that are splices in this creature?

7-5 In one paper we read the following:

Normal segregations from heteroallelic diploids of the type $a1+/+a2$ yield two $a1+$ spores and two $+a2$ spores in each tetrad. Yet, prototrophic or revertant spores $(++)$ occur among the meiotic products of such heteroallelic diploids. For the most part, these wild type recombinants arise as gene conversions, or $3+:1a$ segregations for either parental allele, and to a much lesser extent from reciprocal recombination between the input alleles. Prototrophs, it is generally assumed, represent the consequence of some recombinational event, and their meiotic frequency among random spores has been widely utilized as a measure of the genetic distance between the mutant sites. On the basis of a metric that is essentially an index of non-reciprocal recombination, reasonably consistent genetic fine-structure maps have been elaborated. Clearly, this is a paradoxical situation. [S. Fogel, D. D. Hurst, and R. K. Mortimer (1971) in *Stadler Genetics Symposia, Vols. 1–2*, G. Kimber and F. P. Redei, eds., Univ. of Missouri Agr. Exp. Stn., Columbia, p. 89]

If you can find the paradox, explain it to a member of your family.

7-6 In the same paper we read:

. . . the double site symmetrical conversions [i.e., co-conversions] taken alone provide a partial explanation for . . . map expansion . . . . Mutants in close proximity to each other will typically experience co-conversion. Hence, with prototrophic spore frequency as the metric of "distance" between the sites, we assign a value to the interval that is too low . . . . For longer intervals this "error" is relatively less than that for short intervals. Upon summing the length of such short intervals we should consistently find that the sum is less than the observed recombination frequency for the two termini . . . .

The above assertion is accurate only under special conditions. What are those special conditions?

NOTES

1. M. B. Mitchell (1955) *PNAS* **41**:215.
2. S. Fogel, D. D. Hurst, and R. K. Mortimer (1971) in *Stadler Genetics Symposia, Vols. 1–2*, G. Kimber and F. P. Redei, eds., University of Missouri Agriculture Experiment Station, Columbia, p. 89; C. W. Lawrence, F. Sherman, M. Jackson, and R. A. Gilmore (1975) *G* **81**:615.
3. C. W. Lawrence, F. Sherman, M. Jackson, and R. A. Gilmore (1975) *G* **81**:615; G. R. Fink and C. A. Styles (1974) *G* **77**:231.
4. G. Leblon (1972) *MGG* **115**:36; J-L. Rossignol (1969) *G* **63**:795.
5. G. Leblon (1972) *MGG* **115**:36.
6. Ibid.
7. G. Leblon and J-L. Rossignol (1973) *MGG* **122**:165.
8. S. Fogel and R. K. Mortimer personal communication.
9. J. L. Rossignol (1969) *G* **63**:795.
10. P. Lissouba (1961) *ASNBBV* **44**:641.
11. C. W. Moore and F. Sherman (1977) *G* **85**:1.
12. B. C. Lamb (1975) *MGG* **137**:305.
13. R. C. Lewontin and J. L. Hubby (1966) *G* **54**:595; R. C. Lewontin (1974) *The Genetic Basis of Evolutionary Change*, Columbia University Press.

8

Conversion and Crossing-Over

Conversion and crossing-over in fungi are related events. Up to half of the conversion asci are tetratype, and the converted chromatid is involved in the exchange. In *Ascobolus* the details of the relationship can be understood in terms of mismatch correction of splatches, which seem to be more or less reciprocal. Conversion in yeast may not occur via mismatch correction of heteroduplex splatches. If it does occur that way, the splatches must be nonreciprocal.

FORMAL RELATIONSHIPS BETWEEN CONVERSION AND CROSSING-OVER

In Chapter 7 we stated that in fungi nonreciprocal recombination (conversion) is associated with reciprocal recombination (crossing-over) of flanking markers. The details of that relationship provide provocative data for speculation on the mechanisms of recombination in eukaryotes. Since some of the central features of conversion differ among fungi, it is no surprise that the relationships of conversion and crossing-over differ as well. We shall first state those generalizations that apply to all or most of the studied fungi, and then look at some individual cases.

Crosses that reveal the relationship between nonreciprocal and reciprocal recombination have the following features.[1] Gene B is marked at one or more sites at which conversions are monitored. Genes A and C flank B closely but at a distance great enough to ensure that they do not co-convert with B. D is a marked gene whose presence permits the assessment of chiasma (and chromatid) interference. Thus, for example:

a	$b, b_2{}^+$	c	d

$\times$

a^+	$b, {}^+ b_2$	c^+	d^+

From such crosses the following results appear to have general applicability:

1. Conversion at B (either of site 1, site 2, or co-conversion) is correlated with crossing-over of A and C. Among B convertant tetrads the crossing-over frequency for A and C is higher than in the unselected population. This correlation applies to both conversions with postmeiotic segregation (5:3) and those without (6:2).

2. In a tetrad convertant at B and recombined for A and C, only two of the four chromatids are recombined, as is expected for close markers (Chapter 3). The two that are not recombined for A and C are parental genotype at B; they show no sign of involvement in the conversion. For example, in a tetrad convertant at 1 and recombined for A and C, typical configurations for markers in the four spore pairs of a tetrad are:

6:2 at site 1	5:3 at site 1
$a\ b_1\ b_2{}^+c$	$a\ b_1\ b_2{}^+c$
$a\ b_1\ b_2{}^+c$	$a\ b_1\ b_2{}^+c$
$a^+b_1{}^+b_2{}^+c$	$a^+b_1\ b_2{}^+c$
$a^+b_1{}^+b_2{}^+c$	$a^+b_1{}^+b_2{}^+c$
$a\ b_1{}^+b_2\ c^+$	$a\ b_1{}^+b_2\ c^+$
$a\ b_1{}^+b_2\ c^+$	$a\ b_1{}^+b_2\ c^+$
$a^+b_1{}^+b_2\ c^+$	$a^+b_1{}^+b_2\ c^+$
$a^+b_1{}^+b_2\ c^+$	$a^+b_1{}^+b_2\ c^+$

This correlated simultaneous involvement of a chromatid in crossing-over and conversion argues that reciprocal and nonreciprocal recombination in fungi are two aspects of a single process, and it is on that hopeful note that we shall proceed. In Chapter 5 we examined the evidence for splices in fungi. Many of the postmeiotically segregating spores that were recombinant for flanking markers demonstrably originated from 5:3 conversion asci; aberrant 4:4 tetrads provide others in some fungi.

3. Half or more of the asci convertant at B are parental for A and C. This tells us that conversion can occur without recombination of flanking markers. Thus, if conversion and crossing-over are part of the same process, they are not inseparable parts. Those 5:3 asci that retain parental configuration of flanking markers are a primary source of the spores we used in Chapter 5 to argue for patches in fungi (others derive from aberrant 4:4 types). Those asci that show conversion at site 1 or 2 to yield $b_1{}^+ b_2{}^+$ while retaining the parental configuration of A and C are a major source of the spores we used in Chapter 6 to illustrate localized negative interference in fungi.

4. Conversion at B does not interfere with crossing-over in the CD interval unless the conversion is accompanied by crossing-over of A and C. Here we recognize a parallel with the relationship of splatches to interference as described in Chapter 5; splices interfere with other splices, patches do not. This parallelism arises from the fact that 5:3 and 3:5 conversion asci are the primary source of postmeiotically segregating spores, which reveal splices and patches.

A multitude of models has been marketed in response to the challenge posed by the relationships between conversion, postmeiotic segregation, LNI, crossing-over, and interference. One of the first, the Holliday model, has achieved the status of Hypothesis Emeritus. I suspect that it is not fully correct, but it is efficient; it explains a lot (not all) with a few postulates (perhaps too few). Because of these virtues, it has served as a target against which the fungal geneticist has aimed his crosses. It will provide a temporary framework for our discussions; facts will be interesting as they support or contradict that framework.

THE HOLLIDAY MODEL

At this stage the Holliday model[2] is best stated in terms of its products. Later, when we grapple with molecular mechanisms, we shall specify the postulated intermediates.

1. Splatches of variable length are formed reciprocally from fixed starting points, of which there are many per chromosome but probably fewer than one per gene.

2. Patches and splices are equally frequent and have the same length distribution. Points 1 and 2 are summarized in this diagram:

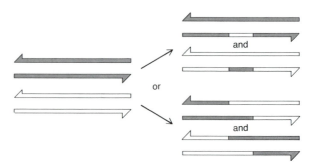

3. Heteroduplexes are corrected to mutant or wild type with an efficiency and direction that varies with the nature of the marker and the peculiarities of the correcting enzymes of the fungus in question.

4. Splice formation causes chiasma interference, while patch formation does not (this feature was not in the original formulation, but should have been).

You might recognize the model; you met a simple form of it in Problems 7-2 and 7-3. The model has one obviously slippery feature; it ascribes conversion to a reaction, correction, which proceeds by totally unspecified rules. Except for that built-in vagueness, the model is admirably explicit. Some of the testable features of the model that have been found wanting are the following.

1. Conversion should be accompanied by recombination of close flanking markers in half the cases. In fact, however, an excess of recombinants over parental types is expected since patch formation sets up no interference while splice formation does. Thus, events independent of a conversion-causing patch can recombine A and C, putting them into the category recombined for flanking markers. Chiasma interference prevents the reverse reaction. However, in many cases it has been observed that *half* or more of the convertants are parental for flanking markers.

2. Correction should be inefficient for a site at which a large fraction of the convertants are 5:3. In these cases one might expect to find 4:4 aberrants in quantity. Although 4:4 aberrants are reported for *Sordaria*[3] and *Ascobolus*,[4] their frequencies in yeast[5] and *Neurospora*[6] are below expectation.

3. Some evidence of splatch formation should be found on two chromatids in at least some conversion asci. In yeast, as we shall see, only one chromatid manifests splatch symptoms, such as postmeiotic segregation (see item 2) or LNI.

4. Since patches and splices have the same length distribution and both are reciprocal, conversion behavior at B should be the same when recombination of flanking markers accompanies the conversion event and when it does not. The *Ascobolus* experiments cited in Chapter 7 in support of heteroduplex correction are quantitatively compatible with reciprocal splatch formation in the Holliday model. As we shall see below, however, more sensitive experiments (in a different *Ascobolus* strain and at a different genetic locus) reveal that splatch formation is not invariably reciprocal.

NONRECIPROCAL PATCHES

Conversion asci are of two sorts with respect to flanking markers: parental or recombinant. In the Holliday model the rules governing conversion are independent of the state of the flanking markers. In particular, splatch formation is reciprocal, and this reciprocality should apply to both classes of conversion asci. The cross described below argues that *Ascobolus* deviates from this prediction; splice formation, which recombines flanking markers, may be reciprocal, but patch formation, which leaves them parental, is probably not. The cross was of the following form:[7]

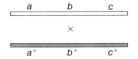

The rare 5:3 asci at B were selected for study. They were classified as to whether they were parental ditype (PD) or tetratype (T) for the flanking markers (A and C were sufficiently close that nonparental ditype (NPD) tetrads were not expected). Among the PDs, two sorts are expected on the Holliday model. The case of $5^m:3^+$ is illustrated in Figure 8-1. Essentially all of the PD tetrads were type 1. The essence of type 1 tetrads, in contrast to type 2, is that three of the four chromatids retain their parental genotype with respect to all three marker loci. Perhaps compatibility with the Holliday model can be argued (try it!), but an economical interpretation is surely that patches (at this locus in this creature) are formed nonreciprocally. (Similar experiments in *Sordaria*[8] revealed type 2 tetrads in quantity, although type 1 were in excess. On the other hand, extensive work in yeast, which we shall describe later in this chapter, strongly supports a picture of nonreciprocal patch formation.)

$5^m:3^+$ asci recombinant for A and C arise in two ways. Reciprocal splice formation would produce type 3 tetrads (Figure 8-2). Type 3 tetrads may

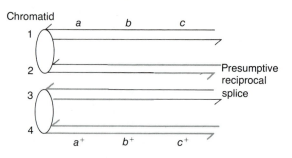

Heteroduplex
correction on:

chromatid 2 | chromatid 3

Type 3a	Type 3b
$a\ b\ c$	$a\ b\ c$
$a\ b\ c$	$a\ b\ c$
$a\ b\ c^+$	$a\ b\ c^+$
$a\ b\ c^+$	$a\ b^+\ c^+$
$a^+\ b\ c$	$a^+\ b\ c$
$a^+\ b^+\ c$	$a^+\ b\ c$
$a^+\ b^+\ c^+$	$a^+\ b^+\ c^+$
$a^+\ b^+\ c^+$	$a^+\ b^+\ c^+$

FIGURE 8-1
The origin of Type 1 and Type 2 tetrads by heteroduplex correction following reciprocal patch formation.

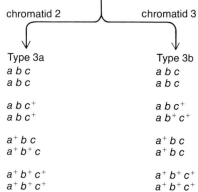

Heteroduplex
correction
to b on:

chromatid 2 | chromatid 3

Type 1	Type 2
$a\ b\ c$	$a\ b\ c$
$a\ b\ c$	$a\ b\ c$
$a\ b\ c$	$a\ b\ c$
$a\ b\ c$	$a\ b^+\ c$
$a^+\ b\ c^+$	$a^+\ b\ c^+$
$a^+\ b^+\ c^+$	$a^+\ b\ c^+$
$a^+\ b^+\ c^+$	$a^+\ b^+\ c^+$
$a^+\ b^+\ c^+$	$a^+\ b^+\ c^+$

FIGURE 8-2
The origin of Type 3a and 3b tetrads by heteroduplex correction following reciprocal splice formation.

also arise by exchanges occurring in the *AC* interval among patch tetrads (recall that patches do not interfere with chromosome exchanges). These exchanges steal from type 1 tetrads to make type 3 when the heteroduplex chromatid is involved in the exchange:

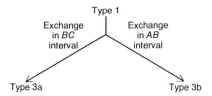

When type 1 tetrads are lost due to exchange not involving the heteroduplex chromatid, they give rise to type 4 asci (Figure 8-3). If the exchange that steals from type 1 tetrads really shows no chromatid interference (either positive or negative) vis-a-vis the patch, then type 1 tetrads will become type 3 and type 4 equally. Then, the observed excess numbers of type 3 over type 4 are the ones that arise by splices. This excess is significant, providing

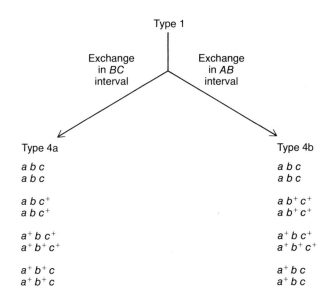

FIGURE 8-3
The origin of Type 4a and 4b tetrads from exchanges occurring in a tetrad that would otherwise have been Type 1.

as good evidence for splices as we have in fungi. Since the model that we used to generate type 3 involved reciprocal splice formation followed by correction of one or the other heteroduplex, the outcome is clearly compatible with reciprocal splices. However, compatibility does not imply support; type 3 tetrads might arise without correction in the event that a splice is formed on only one of the recombinant chromatids. Thus, while it does offer evidence that patch formation need not be reciprocal, the experiment leaves open the matter for the case of splices.

NONRECIPROCAL SPLATCH MODELS

The strong possibility that patches are not (always) reciprocal leads to the following minimally modified model for patch formation (where the broken-line segment of the dark chromatid signifies newly synthesized DNA):

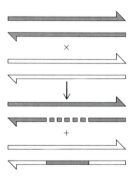

Likewise, the possibility, which we have raised and will address further below, that splices, too, need not be reciprocal leads to the following minimally modified model for splice formation:

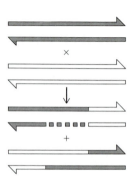

Note that both models offer a basis for conversion in addition to correction of mismatches. In each case, DNA synthesis provides an additional stretch of single chain of one type while the corresponding stretch of the other type has been lost in the shuffle.

Recent years have seen heroic analyses in yeast of unselected asci, i.e., a full genotyping of all products in large numbers of asci not selected for their relevance to any hypothesis.[9] The single outstanding conclusion from these studies is that splatch formation shows no signs whatever of reciprocality, and the Holliday model is forced to yield to models in which at least part of the deviations from 4:4 segregation are based on syntheses and losses other than those resulting from heteroduplex correction. The two key observations are the following. (1) 5:3 asci are frequent for some sites, but aberrant 4:4 asci do not occur except at the low frequency expected for completely independent splatches. (2) Certain asci that are expected from correction of reciprocal patches do not occur. In particular, the following two types are not found, one of which is the type 2 ascus missing in *Ascobolus*:

5:3		Normal 4:4
$a\ b\ c$	$a\ b\ c$	$a\ b\ c$
$a\ b\ c$	$a\ b\ c$	$a\ b\ c$
$a\ b\ c$	$a\ b^+c$	$a\ b^+c$
$a\ b^+c$	$a\ b^+c$	$a\ b^+c$
$a^+b\ c^+$	$a^+b\ c^+$	$a^+b\ c^+$
$a^+b\ c^+$	$a^+b^+c^+$	$a^+b\ c^+$
$a^+b^+c^+$	$a^+b^+c^+$	$a^+b^+c^+$
$a^+b^+c^+$	$a^+b^+c^+$	$a^+b^+c^+$

The failure to find any evidence of reciprocality in splatch formation in yeast deposes the Holliday model in that area and substitutes as a minimal modification the schemes for patch and splice formation presented above. (Holliday defends reciprocal splatch models against this evidence. We will look at the essence of his defense in Chapter 11.)

NONCORRECTION MODELS

In Holliday's scheme, 5:3 asci are explained as a result of correction on one of two reciprocal splatches. In the minimally modified models, 5:3 asci arise as a result of splatch formation on only one of two chromatids; 6:2 asci are explained as the result of correction of that single splatch. The important role of correction in *Ascobolus* has been argued above. For yeast, however, similar evidence for correction is lacking (though map expansion has been reported[10]). This situation has encouraged the formalization of models in which at no step is gene conversion due to correction of hetero-duplexes. The details of some of these schemes will be examined in Chapter 11. Their essence is summed up by an overall reaction equation:

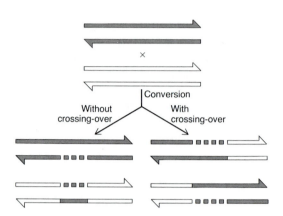

In this example a stretch of white chromatid has been lost, and the loss has been compensated by a semiconservative replication of a stretch of a black chromatid. As drawn, the scheme makes no provision for heteroduplexes, since it is the intention of the scheme to deny their role in conversion. Their reality is easily accomodated, however, by allowing splices at the junctions between black and white, like this:

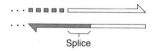

Note that the model lacks patches, calling instead on paired splices. In yeast, different markers manifest different ratios of 5:3 to 6:2 asci. If, as we may imagine, this is not due to differing efficiency of correction, then it must be due to their differing probabilities of being incorporated into a splice. This could be due to marker effects exerted on splice formation or to varying probabilities of splice formation along the chromosome. In these noncorrection models for conversion, the various manifestations of polarity result in much the same way that they do in correction models. They can be ascribed either to fixed starting points (for replication) or to fixed pairing segments, as we shall describe in Chapter 11.

The nonreciprocal splatch models for conversion can be stretched in either of two directions to cover parity in yeast. Either we can retreat into our ignorance of correction enzymes, claiming for them the properties needed, or we can tie the correction process to the formation of the splatch. If splatch formation is asymmetric in a way that allows a distinction between "donor" and "recipient" chains, a possible basis for parity exists. We suppose that correction enzymes can make the hypothesized distinction and perform all correction reactions on the recipient. Whatever the nature of the mismatch, parity would result.

Up to this point I have dealt with recombination in as formal a fashion as I found possible. My aim was to expose the facts to be explained within a framework tight enough to corral them. Now that we have seen what is to be explained, we can depart from formalisms and venture into the chemistry of DNA in a search for mechanisms.

PROBLEMS

8-1 The Holliday model supposes that all conversion is the result of splatch formation followed by mismatch correction. If only a small fraction of the aberrant asci in a particular creature (or at a particular locus) manifest postmeiotic segregation, the Holliday model explains this as a result of highly efficient correction of heteroduplexes. What do you think William of Occam would have thought of this aspect of Holliday's model?

8-2 Consider the cross $abc \times a^+b^+c^+$, where A and C are close to B but far enough from it to escape co-conversion.

(a) Write down the genotype of the ascus that results from reciprocal patch formation at B ignoring independent coincidental recombination events. Allow no mismatch correction.

(b) Correct the mismatches in the ascus of part a. Write down all asci that can result from correction of one or both heteroduplexes. Identify those asci whose absence in yeast argues that patch formation is nonreciprocal.

8-3 In recombination calculations one can be fooled by flying factors of two. They arise from the two-ness of the interaction between chromatids, from the two-ness of the Watson-Crick duplex, and elsewhere. Here is a case of a flying factor to think about. Why does the slope in Figure A-6 in the appendix change by a factor of two at $D/L = 1$ for the case of $a = 0.5$? Got it? Good. Two authors[11] derive the model graphed in Figure A-6 assuming the rigorous validity of Holliday's model, i.e., that splatches are exactly reciprocal and that splices and patches are equally frequent. Then they say ". . . the prediction of an initial slope depends on the formation of [splatches] reciprocally on both participating chromatids . . . . If [splatch] formation is *always* nonreciprocal [and splices and patches are equally frequent], the initial and final slopes should be the same." Decide whether this contention is true or false.

8-4 In one experiment, separate conversions failed to show interference with each other. However, the subclass of conversions that recombined flanking markers did interfere with each other ($S < 1$). These observations led to the proposition that conversions are "laid down" along bivalents at random, then "... some signal between adjacent recombinogenic events must exist to constrain that they are usually of opposite types."[12] In order to formalize this model, let us take it one step further than the authors did and state that adjacent recombinant events are always of opposite types.

(a) Write a mapping function for this model. Limit the function to distances that are outside the range of LNI. Graph your function along with Haldane's Function in order to visualize the positive interference in your function.

(b) For your function, graph S versus $2R_1$ (which is not tedious if you have access to a computer or a programmable calculator).

(c) In yeast, about half the conversions *do* recombine flanking markers. In *Neurospora*, however, the fraction of conversions associated with chromosome exchange is more like one-third. Write a mapping function, valid for markers that are not very close, on the assumption that recombination events are Poisson-distributed among bivalents and that each *third* event exchanges flanking material. Convince yourself (graph the function, if need be) that this function has stronger interference than that of part *a*.

For different fungi, it would be interesting to compare the intensity of positive interference with the fraction of conversions that recombine flanking markers. Perhaps such data will someday become available.

The data[13] that prompted this model were not quite adequate to rule out an alternative. The alternative is that events involving exchange of flanking material are laid down with positive interference according to *some* rule, and then events of the other kind are sprinkled at random among them.

8-5 The model of recombinogenic events with alternating fates (Problem 8-4) has other interesting properties. (1) it predicts interference between events each of which fail to have an associated chromosome exchange, and (2) it predicts negative interference between the two different kinds of events. Point 1 is proven by analogy with the argument of Problem 8-4. In this problem we look at point 2.

Consider the cross $abcd \times a^+b^+c^+d^+$, where the R values are all pretty small, so we do not have to worry about more than one event per interval. I contend that among tetrads convertant at B but parental ditype for A and C, the fractions that are tetratype for C and D will be higher than in the overall population. By what factor will it be higher?

8-6 Devise a method for detecting aberrant 4:4 segregation in a fungus with unordered eight-spore asci.

NOTES

1. M. B. Mitchell (1955) *PNAS* **41**:215; S. Fogel and D. D. Hurst (1967) *G* **57**:455.

2. R. Holliday (1964) *GR* **5**:282.

3. Y. Kitani, L. S. Olive, and A. S. El-Ani (1962) *AJB* **49**:697.

4. G. Leblon and J-L Rossignol (1973) *MGG* **122**:165.

5. S. Fogel and R. K. Mortimer (1970) *MGG* **109**:177; (1974) *G* **77** (Suppl.):52.

6. D. R. Stadler and A. M. Towe (1963) *G* **48**:1323.

7. D. R. Stadler and A. M. Towe (1971) *G* **68**:401.

8. Y. Kitani and L. S. Olive (1969) *G* **62**:23.

9. S. Fogel, D. D. Hurst, and R. K. Mortimer (1971) in *Stadler Genetics Symposia*, *Vols. 1–2*, G. Kimber and F. P. Redei, eds., University of Missouri Agriculture Experiment Station, Columbia, p. 89; S. Fogel, R. Mortimer, K. Lusnak, and F. Tavares, *CSHSQB*, **43**, in press. S. Fogel and D. D. Hurst (1967) *G* **57**:455.

10. M. S. Esposito (1968) *G* **58**:507. E. W. Jones (1972) *G* **70**:233. R. Holliday (1974) *G* **78**:273.

11. J. R. S. Fincham and R. Holliday (1970) *MGG* **109**:309.

12. R. K. Mortimer and S. Fogel (1974) in *Mechanisms in Recombination* R. F. Grell, ed., New York: Plenum, p. 263.

13. Ibid.

9

Recombination Chemistry

The DNA duplex is maintained by the weak forces of hydrogen bonding and base stacking, which allow appreciable flexibility. Unstacking, annealing, strand migration, and supercoiling are of potential interest to recombination. Proteins of interest include exo- and endonucleases of various sorts, unwinding enzymes, single-chain binding proteins, and polymerases. For some of these, the genes that inform the proteins have been identified and the effects on recombination of mutations in the gene described. Other genes of interest have been identified by the effects of their mutants on recombination. Mutations that create (or abolish) sites that exalt recombination in their neighborhood may signal the unique points responsible for the phenomena that have been previously explained by the concepts of fixed pairing regions and fixed starting points for splatches. Some reaction sequences that have been demonstrated in vitro may reflect in vivo steps in recombination.

GENERAL CONSIDERATIONS

Our goal in this book is to approach an understanding of recombination in chemical terms. Complementary avenues to this goal must be traveled. So far in this book we have been concerned with the genetic structure of recombinant molecules; we have discussed the rules governing the genetic (informational) contributions from each of the recombining chromosomes. We have also considered, in broad outline, the material contributions from the interacting chromosomes. In both cases we were describing the ultimate products of recombination reactions without worrying about the chemical plausibility of possible routes for the formation of those products. That is what we shall start to do now.

In this chapter we shall look at the chemistry of DNA with respect to those features that are known to be, or threaten to be, immediately relevant to genetic recombination. The kinds of things we must consider are implied, at least partly, by the previous chapters. Phosphodiester bonds must be broken, perhaps at specific places; single-chain stretches must be exposed; exposed complementary chains from different chromosomes must form Watson-Crick duplexes with each other; some DNA must be synthesized; chain continuity must be secured; mismatches must be recognized and corrected. In considering the chemistry of these reactions, we shall divide our task into several parts.

1. The physicochemical properties of DNA itself suggest ways we should think about recombination.
2. The kinds of enzyme reactions and other DNA-protein interactions already catalogued suggest plausible steps in recombination.
3. In a number of cases the involvement of a particular protein in recombination is directly implicated by recombination-deficient phenotypes of strains lacking the protein.
4. Treatments of DNA that result in altered recombinational behavior suggest possible intermediates in recombination.
5. Recombination-like DNA interactions in vitro support certain propositions regarding mechanisms involved in vivo.

In Chapters 10 and 11, we shall unite our genetics with our newly acquired chemistry as we consider detailed models for various pathways of recombination.

PHYSICAL CHEMISTRY OF DNA

The duplex structure of DNA is maintained by hydrophobic interactions between the parallel base pairs and by hydrogen bonding between the members of each pair. Both of these forces are weak with consequences of possible importance to recombination. These include breathing, melting, annealing, chain migration, and supercoiling.

In a long DNA molecule, any moment finds occasional short stretches of the duplex dissociated (denatured, unstacked). Reassociation occurs quickly, however, because the two chains are held close to each other (i.e., they are kept at high local concentration) by regions of intact duplex above and below.

We may diagram a breathing duplex like this:[1]

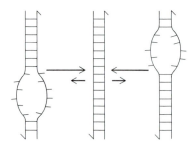

The return of a breathing region to the Watson-Crick configuration is promoted by the well-stacked regions above and below it. A breath taken at the end of a duplex is a deeper one; return to the stacked state is aided only by the region below.

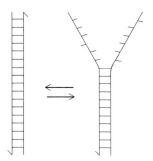

Linear duplexes are usually somewhat denatured at their tips. The possible importance of breathing for recombination lies in the possibilities afforded for the recognition of base sequence. In duplex DNA the distinctive pattern of H-bond donor and acceptor groups of each chain is hidden by H-bonding with the complement. Disruption of H-bonds at physiological conditions, i.e., breathing, might conceivably help such outside agents as enzymes or single-chain polynucleotides to "see" a bit of the base sequence.

   As temperature (or pH) is raised (or ion concentration lowered), the fraction of any duplex that is transiently dissociated increases. As conditions become more extreme, some stretches of the molecule are open essentially all of the time; i.e., they are "melted." AT-rich regions are the first to be melted. With a slight further rise in temperature, the melted regions extend from their origins in AT-rich regions until they unite, and the duplex becomes completely unstacked.[2] If the DNA molecule is linear, or is a nicked circle (a circle with at least one broken phosphodiester bond), the complementary chains separate.

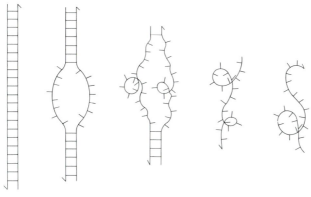

Increasing temperature or pH $\longrightarrow$

In covalently closed circles, of course, the two chains remain entangled and twisted about each other as many times as they were in the Watson-Crick state prior to unstacking. Complete melting may be without physiological significance. Local melting or "deep breathing", however, may play a role in splatch formation.

Completely melted DNA can be returned to its original Watson-Crick configuration; melting, even complete melting, is reversible.[3] The degree of reversibility of the melting reaction is independent of the rate or temperature at which melting was achieved but is strongly dependent on the rate at which the original conditions of temperature (or pH) are restored. Rapid cooling "traps" the chains in configurations in which some base pairing and stacking is achieved by fortuitous association between quasi-complementary regions, usually of the same chain. These "incorrect" renaturations preempt the opportunities for achieving fully correct, extramolecular associations. Slow cooling allows chains to avoid or escape from traps and to stop their "searching" only when Watson-Crick status is restored. Trapped DNA can be saved by heating, to restore the fully melted state, followed by slow cooling. This process is superficially akin to the annealing of steel or glass to make it less brittle. Hence, renatured DNA is "annealed" DNA.

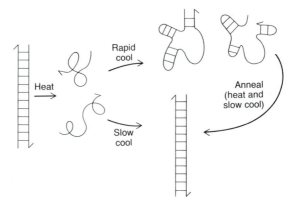

Splatch formation almost certainly involves annealing. The in vivo reality of the reaction has been illustrated in the case of association of the sticky ends of the λ chromosome following DNA injection (Chapter 4).

The weakness of the forces holding complementary chains together permits reversible reassociations at physiologic conditions.[4] These reactions can be illustrated by two chains paired with a sole complementary chain.

Each micro-step in the reaction involves the breaking and making of one Watson-Crick base pair, so all of the possible states are equally likely. Furthermore, the weakness of the individual bonds involved means that the activation energy is low. Thus, the reactions diagrammed are rapid and reversible at physiologic conditions. This process of branch migration probably plays a role in splatch formation.

A variety of proteins bind to DNA and in so doing change slightly the pitch of the helix in their locale. The changed pitch necessitates a local reduction in the number of twists per base pair (normally about 0.1), and if the helix is free to rotate about its axis, these twists are lost from the duplex. Such is the case with linear duplexes (as long as they are not tied to something) or nicked circles. The linear duplex may in the course of events become circularized (or otherwise constrained) or the nicked circle may become covalently closed. This DNA, extracted and deproteinized (without nicking or breaking), is topologically unable to adjust its number of twists to the original value. The most stable structure for such a molecule is a compromise one in which supercoils permit increased H-bond and stacking interactions.[5] The supercoiled structure is not as energetically favored as the simple Watson-Crick state, so that introduction of a single nick instantly results in loss of the supercoils, and the resulting molecule is said to be "relaxed."

Relaxation can be achieved by replacing the protein lost in extraction. Interestingly, relaxation can also be achieved by adding pieces of homologous single-chain DNA, which then become "annealed" to the supercoil at physiologic conditions.[6] This ability of supercoiled DNA to bind homologous single-chain DNA may also play a role in splatch formation.

A role for supercoiling has been demonstrated in the site-specific recombination promoted by the Int system of λ.[7] In this case, which is described in further detail in Chapter 10, supercoiling can be achieved in vitro through the catalytic action of a purified protein. A large fraction of the in vivo supercoiling in *E. coli* is demonstrably due to the catalytic activity of such a "gyrase" enzyme.[8]

The efficient renaturation of DNA in vitro requires *slow* cooling to prevent the entrapment of chains in quasi-complementary interactions. In vivo annealing appears to be protein catalyzed. An "annealing protein" binds to single-chain DNA to prevent intramolecular quasi-complementary renaturation and thereby facilitates proper renaturation.[9]

BIOCHEMISTRY OF DNA

The linear continuity of DNA chains is provided by phosphodiester bridges between adjacent deoxyribose residues. The enzymatic making and breaking of these bonds are certain to be involved in recombination.

Enzymes that cut DNA chains without regard to the proximity of the ends are called *endonucleases*. The endonucleases vary in their action in several respects. Some act only on duplex DNA; these are subdivided into those that cut but one of the two chains ("nickases") and others that cut through both chains ("chopases"). Other endonucleases act only on single-chain DNA. (Of course, the enzyme is sensitive to the number of chains only in its immediate neighborhood. Thus, a single-chain endonuclease will cut a single-chain region in a predominantly duplex molecule.) In addition to the two major categories, there are other types of endonucleases. Some cut such that the phosphate group is left on the 3'OH group or on the 5'OH of the neighboring sugar. This may be important in determining which enzyme can operate on the product in the subsequent step in a reaction series leading to recombination. Some enzymes are insensitive to base sequence, others show rather vague preferences, and still others are specific for a single sequence.

Enzymes that cut up DNA chains progressively from their ends are called *exonucleases*. These, too, vary in ways that may be relevant to recombination. Some digest only single-chain DNA, generally in a particular chemical direction. Others work on duplex DNA, digesting in only one chemical direction:

Other double-chain exonucleases digest the chains in both chemical directions:

Some exonucleases can initiate digestion at a nick:

Other duplex-specific exonucleases require a gap in order to initiate digestion from within:

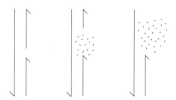

Enzymes involved in DNA synthesis[10] are candidates for roles in recombination. Classification of these enzymes is more complex than that of the nucleases. A major division distinguishes between enzymes required for chain elongation (polymerases) and chain initiation. The polymerases require a template and a primer, as well as nucleoside triphosphates, and can elongate chains in only one chemical direction. The template is single-chain DNA; the primer is a piece of complementary polynucleotide in Watson-Crick pairing with the template:

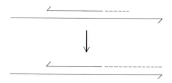

The polymerase adds nucleotides to the 3'OH end of the primer like this (where the broken line signifies new DNA):

In the above diagram chain extension catalyzed by polymerase is converting a single-chain region into a duplex. Chain extension can create single chain regions, too, in a "chain displacement" reaction:[11]

DNA synthesis at a replication fork requires chain extension, as diagrammed above, and, in addition, repeated chain initiations. The initiations provide primers for the synthesis of that chain whose chemical direction for elongation is opposite to its topological growth direction:

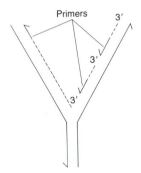

The primers may be short stretches of RNA. The primer plus the piece of nascent chain attached to it is called an *Okazaki piece*; Okazaki pieces are, very roughly, one gene long. Completion of the new chain on the right arm of the fork in the above diagram requires polymerase action to elongate the chains, RNase action to digest primers, and ligase action (see below) to provide final covalent continuity of the chain. The needed RNase activity is, conveniently, a property of the same protein that has the polymerase activity.[12] Nicks in duplex DNA can be closed by the enzyme ligase.[13]

Endonucleases that attack single-chain DNA will cut an unstacked region in an otherwise duplex molecule:

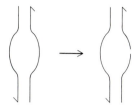

Such unstacked regions exist in heteroduplexes involving gross mutations. Some single-chain endonucleases are responsive even to the slightly unstacked region resulting from a single-base mismatch.

All of the classes of enzymes we have described are useful in modeling recombination, and some of them have been directly implicated by studies like those described below.

MUTATIONS AFFECTING RECOMBINATION

Many mutations are known that have consequences for recombination. (We are not now talking about mutations that perturb the process locally through marker effects, but of mutations that have widespread consequences.) From this array we shall pick for description only those that meet at least one of two criteria: they must be strong candidates for an immediate involvement in exchange and/or correction, or the enzyme or substrate affected must be known or suspected. By such a selection, we are surely slighting mutations that someday soon will have a lot to offer.

Modern studies of recombination-deficient mutants were begun with *E. coli*, the bacterium that is host to the most intensively studied phages.[14] The mutants (*rec*) were identified as F⁻ (female) bacteria that were competent to accept DNA in generalized transduction or in conjugation but were unable to recombine their own chromosomes with it. (They have Rec⁻ phenotype.) The mutants isolated define three genes. The genes $recB^+$ and $recC^+$ cooperate to produce a nuclease;[15] the $recA^+$ product appears to be an endopeptidase as well as to be a single-chain-binding protein.[16] The relevance to recombination of $recA^+$ endopeptidase activity is mysterious. The single-chain-binding activity, on the other hand, can catalyze annealing in vitro (T. Shibata, C. DasGupta, R. Cunningham, and C. Radding, personal communication) and may be fairly presumed to do so in vivo.

The $recBC^+$ nuclease has exonucleolytic activity and has been tagged ExoV. It attacks double- and single-chain DNA, degrading them without regard to their chemical direction. The products of degradation are oligonucleotides, telling us that the enzyme typically cuts phosphodiester bonds about 200 nucleotides from the terminus of the chain. This property of the enzyme prompts a speculation that we may wish to refer to later. Perhaps the site of enzyme action is at the junction of a melted and nonmelted region:

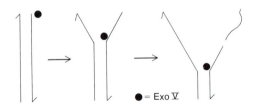

In the above diagram ExoV is shown acting at a duplex terminus. It can act as well from a gap:

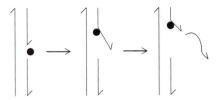

As with other DNA-unwinding enzymes,[17] the unwinding activity of ExoV is ATP-dependent. In the absence of an annealing protein[18] or in low ATP concentration[19] ExoV digests the oligonucleotides to fragments a few nucleotides long. The guess that ExoV acts at the end of a melted region relates its well-established exonuclease activity to its reported endonuclease activity. Perhaps when acting endonucleolytically, ExoV cuts a gapped molecule like this:

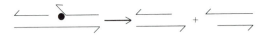

or single chains next to regions of quasi-Watson-Crick structure:

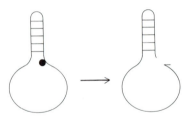

There is circumstantial evidence both for and against this speculation. The *gam* gene of λ informs a protein that binds to and reversibly inactivates ExoV activity. Any condition that results in the reduction of the one activity by *gam*[+] protein results in the same degree of reduction of the other,[20] suggesting that the exo- and endonucleolytic activities are closely related. The counter-evidence is the in vitro properties of ExoV isolated from a certain *E. coli* strain that is mutant in both the *recB* and *recC* genes. The mutant enzyme shows differential temperature sensitivity for the two activities, losing exonuclease activity more rapidly than endonuclease activity with increasing temperature.[21]

The *recA*[+] gene is required for effective recovery of bacteria from the effects of chromosomal damage,[22] for the mutagenic effectiveness of ultraviolet radiation and numerous other DNA-damaging agents,[23] and for the ability of all those agents to induce lytic development of prophages, such as λ.[24] And it is required for all generalized recombination of *E. coli*.[25] About its role in recombination almost nothing is known, but that will not stop us from speculating about it later. Lambda can recombine in *recA* (and *recB/C*) bacteria,[26] revealing thereby that λ has a recombination system (Red) of its own.[27]

The *red α*[+] gene of lambda's Red system informs an exonuclease (Exoλ) specific for double-chain DNA.[28] It removes nucleotides one at a time processively, i.e., without letting go of the DNA between snips, from the 5′ end. Since it is duplex-specific, the enzyme stops when half of the duplex has been digested.

The *red β*[+] gene of λ informs a protein (beta protein) required for operation of the Red system, but is otherwise of unknown activity.[29] It can be found associated with Exoλ in vivo, but the activity of the exonuclease in vitro is uninfluenced by added beta protein.

In Chapter 4 we described an apparent interaction between Red and Rec (to which we shall return in Chapter 10). To a large degree, however, the Red system of λ and the Rec system of its host are independent of each other. Red⁻ λ can recombine in Rec⁺ bacteria, and Red⁺ λ can recombine in Rec⁻ bacteria. This holds no matter which mutations are used to achieve the Rec⁻ or Red⁻ state. The situation with *recB/C* strains of bacteria is more complex, as we shall now see.

Bacteria that are Rec⁻ by virtue of mutation in the *recB* and/or *recC* genes retain a low level of recombination ability. This level is restored more or less to the normal Rec⁺ level by mutation in either of two genes, called *sbcA* and *sbcB* (suppressors of *recBC*). Of these two mutations with similar phenotypes, one (*sbcA*) results in the production of a new exonuclease while the other results in the loss of one![30]

The new exonuclease (Exo III) in *sbcA* strains appears by several criteria to be similar in its mode of action to Exo λ. Apparently, those strains of *E.*

coli that are capable of producing *sbcA* mutants harbor a repressed segment of a lambdoid phage bearing part of the *red* region. On this view, the *sbcA* mutation inactivates the repressor, turning on whatever red^+-like genes are present in the "ghost" segment.

The exonuclease (Exo I) eliminated in *sbcB* strains is (in vitro) a nuclease for single chains and digests from the 3'OH end.

When we grapple with recombination models we shall have to confront two additional complexities in the recombination proficient *recB/C sbc* strains. (1) Lambda, as well as *E. coli*, recombines in these strains. For the case of *E. coli*, $recA^+$ function is absolutely required. For λ, however, *red* (and red^+) phage recombine in *recB/C sbcA* strains even when they are *recA*. (2) The $recF^+$ gene, whose product is unknown, is required for bacterial recombination in both *sbc* strains. However, for λ recombination the $recF^+$ function is required only in the *sbcB* derivatives of *recB/C* strains.[31] It will not be easy to encompass this extraordinary set of observations with a molecular model, but their formal relationships have been neatly summarized (Figure 9-1).

In T4, many mutations involved in DNA metabolism influence the recombination rate. The effects of some of these are not simply consequences of uninteresting changes in the mating regime, such as a change in the length of time between the onset of recombination and the encapsidation of chromosomes, or the lysis of the cells, or changes in the number of chromosomes that can participate in growth in any single infected cell.

Some mutations in T4 decrease recombination while others increase it.[32] Those genes in which mutation has been observed to decrease recombination includes genes *32*, *46* and *47*. Gene *32* specifies the single-chain binding protein of T4. Its roles in recombination might include protection of single-chain regions in recombinational intermediates from destruction by nucleases and annealing of DNA from different molecules to form splatches. Genes *46* and *47* appear to inform an exonuclease whose role in recombination might be to expose single chains, thus facilitating splatch formation.[33] Mutations in these genes prevent the appearance of gaps in T4 DNA blocked in replication as well as the branched molecules described in Chapter 5.

Mutations that increase the number, length, or duration of single-chain gaps all increase recombination frequencies. Such mutations are found in T4's DNA polymerase gene and in its ligase gene, among others. Bacterial recombination is stimulated by defects in the *E. coli* polymerase I and ligase genes, as well.[34] This indication that gapped DNA is recombinogenic is supported by studies on the mechanisms by which DNA-damaging agents increase recombination.

Agents that induce any of a variety of chemical alterations in DNA can increase rate of recombination. Ultraviolet light (UV) is a conveniently applied and well-studied example. The damages induced by UV (notably

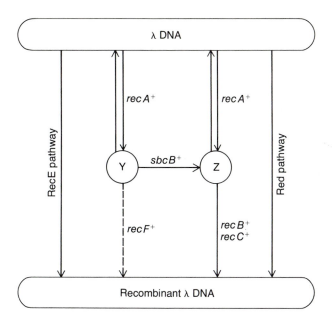

FIGURE 9-1
Lambda can be recombined by any of four sets of enzymes (in
addition to Int) whose relationships are summarized in this
pathway scheme. The dashed arrow from intermediate Y
signifies that the RecF pathway is lightly used when *recB* and
recC gene products are active, is more heavily used in *gam*+
crosses, and is a major Rec route to λ recombination in
recB/C sbcB host cells. (J. R. Gillen [1974] Ph.D. thesis,
University of California, Berkeley.)

cyclobutane dimers of adjacent pyrimidines) lead to the formation of gaps,
and these gaps are presumed to be responsible for the increased recombination.[35]

Ionizing radiations, e.g., X-rays, also stimulate recombination. The
primary damage in this case is a mixture of single-chain and double-chain
breaks. By "double-chain breaks" is meant the induction by a single X-ray
quantum of breaks in both chains of a duplex at essentially the same level in
the molecule so that the duplex is effectively cut into two separate pieces:

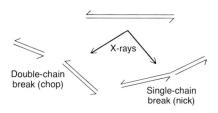

Single-chain breaks induced by X-ray are rarely lethal; they are promptly closed by enzymatic activities in the cell. This prompt repair may annul any opportunity for these nicks to play a role in recombination. The double-strand breaks account for the lethal effects of X-ray.[36] In phage, each break is invariably lethal when an X-rayed particle makes a solo infection. Under conditions of multiple infection, however, some phages show an enhanced ability to survive; a cell multiply infected with dead particles often produces phage.[37] How can fragments from two or more dead phages cooperate to produce viable phages when the fragments from any one of them is insufficient? An obvious candidate is genetic exchange; the fragments could be pieced together by exchanges.

There are two ways of thinking about these exchanges: the exchanges might simply be those that would have occurred even had the infecting phages not been irradiated, or they might occur at the ends of fragments, stimulated in fact by the ends themselves. These two views make one common prediction—the frequency of recombinants from a cross between X-rayed phages should be higher than normal. However, in the first alternative the increase in recombinant frequency is a consequence of selection; those infected cells lucky enough to have many, providentially placed exchanges are the ones that succeed in producing phage progeny. The high recombinant frequency in this case is accompanied by a failure of progeny production by those cells cursed by fate with a low number of exchanges. In the second alternative the increased recombinant production may be observable even when all cells are producing progeny; not only is recombinant *frequency* enhanced, but the *number* of recombinants produced is enhanced by X-rays as well. By this criterion it has been demonstrated for some situations that X-radiation is a true stimulator of recombination. The experiments suggest that it is the double-strand breaks that are recombinogenic.[38] This cannot surprise us in view of the well-established role of T4 ends in recombination, and it reminds us that we had better keep chopped duplexes in mind as possible intermediates in a general model for recombination.

In the fungus *Ustilago*, one endonuclease, DNase I, has been strongly implicated in recombination, probably at a mismatch-stimulated step. Mutants lacking this enzyme show reduced recombination for very close markers. In vitro, the enzyme is active on single-chain DNA. However, it nicks native DNA and nicks duplexes damaged by UV irradiation even faster. The role of the enzyme in conversion is nicely indicated by its action on infectious phage DNA.[39] Naked duplex chromosome of the *B. subtilis* phage SPO1 lose infectivity when treated with *Ustilago* DNase I. Chromosomes that have been denatured and annealed lose infectivity somewhat faster (indicating merely that renaturation is slightly imperfect.) Hetero-duplex DNA made by annealing strands of opposite polarity from two different genotypes of phage loses infectivity conspicuously faster than the control homoduplexes. Thus, the enzyme recognizes and nicks at base mismatches, declaring itself a good candidate for the first enzyme in a mismatch correction pathway involved in gene conversion.

Two genes involved in mismatch correction in *E. coli* appear to have been identified. These are $uvrD^+$ and $uvrE^+$, known to be involved in excision-repair of UV damages. They have been implicated in mismatch correction in two experiments. (1) If artificial heteroduplexes of λ transfect $uvrD$ or $uvrE$ cells of *E. coli*, they give genetically mixed progenies more often than they do when they transfect wild-type *E. coli*.[40] (2) Bromouracil, which induces mutations via a heteroduplex intermediate, is more mutagenic in $uvrD$ and $uvrE$ strains than it is in wild type.[41] Neither of the experiments provides secure evidence that these genes have a role in mismatch correction, but they certainly point to the possibility.

Recombination proteins of a completely unanticipated sort are implied by certain variants of *Neurospora*. Laboratory strains of *N. crassa* are distinguished by single gene differences that have local effects on the rates of recombination. The noted effects are remote from the gene, frequently being detected in chromosomes other than those on which the gene resides. One of the two homozygous types has a high recombination rate in specific remote regions while the other has a low rate in those regions. Surprisingly, the allele conferring the low rate is dominant. This result implies that one of the genetic controls on recombination in fungi is the repression of the action of recombination proteins. If that is the proper explanation, these *rec* mutants of *Neurospora*, though interesting, may do little to elucidate the mechanisms by which recombination is achieved.

In yeast, the recessive mutation *rec4* reduces both the reciprocal and nonreciprocal recombination rate between markers within the *arg4* gene. However, the overall conversion rate for any *arg4* marker remains unchanged, and the fraction of convertants giving postmeiotic segregation remains close to zero. Furthermore, $6^m:2^+$ remain equal to $6^+:2^m$, and the R values for markers flanking *arg4* are the same in *rec4* and rec^+ strains. The reduced recombination frequency (both reciprocal and nonreciprocal) is entirely attributable to a decrease in the probability that a conversion segment will begin or end in the *arg4* gene. This is reflected in an increase (to 100 percent in the fraction of conversions that are co-conversions for the two *arg4* markers in a cross. This mutant looks like a promising one for defining mechanisms of conversion and of crossing-over and the relationships between the two. In Chapter 11 we shall try to accommodate its behavior in a model for recombination in yeast. By the way, *rec4* is not linked to *arg4*, and *arg4* is the only gene of almost a dozen examined whose recombination is influenced by *rec4*!

In *Neurospora*, a genetic element near the *his-3* locus exists in two states; one (*cog*$^+$) confers a high rate of recombination on the *his-3* locus while the other (*cog*$^-$) confers a low rate.[42] These variants differ from the *rec* variants in two respects: the effect on recombination is confined to the neighborhood of *cog*, and the heterozygote *cog*$^+$/*cog*$^-$ has a high recombination rate like that of the *cog*$^+$ homozygote.

Some properties of *cog* suggest that it corresponds to the fixed element in a FSP model. In a cross of the type

a^+	$b_1 b_2{}^+$	*cog*	c^+
a	$b_1{}^+ b_2$	*cog*	c

the majority of B^+ recombinants that retain parental genotype of the flanking markers were a^+c^+ when both parents were *cog*$^-$ but were ac when both were *cog*$^+$.[43] Thus, *cog*$^+$ reverses the recombinational polarity of the *his-3* gene, provoking a high rate of "conversion" for those sites nearest itself ("conversion" is in quotes because the lack of tetrad analysis in these studies makes the use of the word a bit presumptuous). In *cog*$^+$ crosses, as in *cog*$^-$ ones, about half of the *his*$^+$ recombinants are parental and half recombinant for the flanking markers. If we presume splatch formation, then patches and splices appear to be about equally frequent. An interesting property of *cog* is revealed in the cross diagrammed above, when one parent is *cog*$^+$ and the other *cog*$^-$. Among B^+ spores the AC parental genotype associated with the *cog*$^+$ element is more frequent than that associated with *cog*$^-$, regardless of which of the two parents diagrammed in the cross is *cog*$^+$. The immediate impact of this result is obvious; models for recombination in which the two parents play exactly equal roles will not be universally applicable.

Further restrictions on the mechanism by which *cog*$^+$ might stimulate recombination are imposed by the results of a cross in which the *his-3* region, adjacent to *cog*, is split by a translocation.[44] We may diagram the paired chromosomes of the translocation heterozygote at meiosis this way:

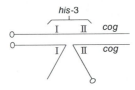

Recombination Chemistry

(I have not bothered to indicate that each chromosome is composed of two chromatids; it would confuse the diagram unnecessarily. The white circles are centromeres.) Recombination between sites in region I as well as between sites in region II was measured in the presence of two of the four possible combinations of *cog* alleles. Recombination in region II was *cog*[+]-stimulated as long as one of the two elements was *cog*[+]. (The lower chromosome was *cog*[+] in the crosses performed.) However, recombination in region I required that the *cog* element on the chromosome bearing the continuous *his-3* region be *cog*[+]. Apparently, stimulation by *cog*[+] involves the transmission of a signal *along* the chromosome. This conclusion will influence our model-making in later chapters.

Studies of *cog* are promising for the insights they may yield on recombination mechanisms; it is sobering, however, to realize that the *cog*[+] and *cog*[-] alleles, both present in standard laboratory strains of *Neurospora*, have for so long been an unrecognized variable in recombination studies. A similar situation characterizes the *rec* mutants of *Neurospora*.[45]

The infectious cycle of λ (Chapter 4) provides several routes for the formation of multimeric chromosomes, which are a prerequisite to encapsidation. In *gam red* strains of λ, only the routes provided by Rec and Int remain, and of these only Rec is sufficient to ensure plaque formation.[46] Thus, *red gam* λ does not make plaques on *recA* bacteria,[47] and only small plaques on *rec*[+] bacteria.[48] Variants that make large plaques on *rec*[+] bacteria are easily selected.[49] They result from the introduction into λ of recombination-stimulating elements similar in many ways to *cog*.[50] These elements, called Chi, arise in λ by one of two routes. First, when λ is grown lytically, the Chi variants arise by a change of one or a few base pairs. They have been detected at four or five sites, indicating that λ carries four (or five) nucleotide sequences within which a modest change creates an active Chi element. Second, when λ is induced from a prophage site on the *E. coli* chromosome, the excision is sometimes faulty, such that a segment of λ is left behind and a segment of *E. coli* chromosome of comparable size is incorporated into λ (Figure 9-2). If the λ is *red gam*, or if it becomes so as a result of a faulty excision, the incorporated piece of *E. coli* DNA may confer a large-plaque phenotype. When it does so, the large plaque is demonstrably a consequence of the presence in the bacterial DNA fragment of *chi*[+]. These elements exist in wild type *E. coli* about once per 10 genes.[51] The properties of Chi are the following (points of similarity with *cog* should be noted): (1) Chi stimulates recombination in its own neighborhood (Figure 9-3); the stimulation falls off with distance, becoming undetectable about 20,000 bases away. (2) When one of the parents in a λ cross is *chi*[+] and the other *chi*[-], the Chi stimulation is strong.[52] (3) Chi stimulates recombination across

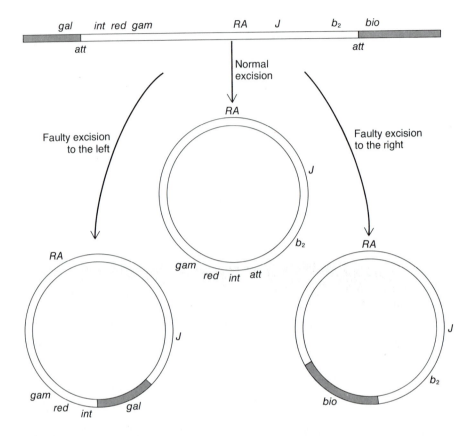

FIGURE 9-2
Formation of λ chromosomes carrying segments of *E. coli* as a result of faulty excision of the prophage.

regions of gross heterology.[53] In each of the crosses below, Chi stimulates recombination in regions indicated by the arrows (the wavy lines indicate the gross heterologies):

Thus, like *cog*, Chi must send a recombinogenic signal along the chromosome.

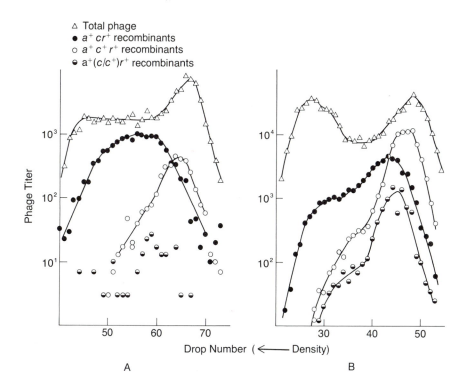

FIGURE 9-3

The density distributions of progenies from replication-blocked crosses in λ. The cross in graph A involves no Chi, while that in graph B has a Chi in each parent near the right end of the chromosome, to the left of gene *R*. The crosses were

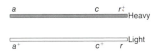

Plaques from total progeny particles are shown as triangles. a^+r^+ recombinants, shown as circles, were either *c* (*filled circles*) or c^+ (*open circles*) depending on whether the exchange occurred left or right of the *c* marker, respectively. Plaques that were mixed for *c* and c^+ (*half-filled circles*) are interpreted as signifying splices. The low frequency of mixed plaques seen in the cross without Chi may either reflect difficulty in detecting them in these phages or indicate splices that are shorter than those stimulated by Chi. (F. W. Stahl and M. M. Stahl [1975] *MGG* **140**:29.)

In general terms, both Chi and *cog* appear to be recognition sites for an enzyme involved in recombination. For Chi, the products of the event include chromosome exchanges (with splices, Figure 9-3). The exchanges are reciprocal in at least a fraction of the cases as judged by physical evidence; an enhanced level of oligomeric circles is detected by sedimentation analysis.[54] It is not known whether Chi stimulates patch formation; since patches do not produce chromosome dimers, patched chromosomes remain unencapsidated unless rescued by an exchange (involving splice formation.) In the presence of Chi, the rescuing splice will usually occur near Chi, obfuscating efforts to detect a patch. For *cog*, postmeiotic segregation has not been reported. However, if we assume that a high frequency of parental arrangement of flanking markers among very close marker recombinants implies patch formation, then splices and patches must be stimulated about equally by *cog*.

Special features of the λ life cycle have made possible a unique demonstration of a correlation between recombination and DNA synthesis. In a replication-blocked cross chromosomes can become dimeric, and thereby maturable, only via recombination. When the cross is *red gam int*, the RecBC pathway accounts for all mature phage production. If the phages involved are *chi*$^+$, then most of the matured phages will have recombined near the Chi site. When such a cross was conducted in a medium with ^{32}P, the resulting λ particles were slightly radioactive. The chromosomes of such particles were cut into specific fragments (by EcoRI restriction enzyme), and the specific activity of the separated fragments determined. The fragment containing Chi was the most radioactive of all.[55]

Chi was discovered in λ subject to recombination by the wild-type Rec system of *E. coli* unaffected by *gam* protein. It was subsequently tested for its ability to stimulate recombination by the Red system of λ and by the Rec pathways revealed by the *scbA* and *scbB* mutations.[56] Chi proved to be without effect in all cases; it is apparently specific to the wild-type *E. coli* system requiring both *recA*$^+$ and *recBC*$^+$ functions for its expression. In Chapter 10 we will attempt to build that provocative fact into a model for Chi-stimulated, Rec-mediated recombination.

A mutation with properties akin to those of Chi and *cog* has been noted in *Schizosaccharomyces pombe*, a fission yeast.[57] The *M26* mutation in the *ade-6* gene of *S. pombe* has a higher rate of conversion than other mutations in that gene, including ones close to it. Furthermore, unlike those other mutations, it converts almost exclusively in one direction, from mutant to wild type. Markers located near *M26* on either side of it tend to co-convert with *M26*. It is unlikely that *M26* is a deletion or addition, since it is an amber mutation, and amber mutations are typically the result of single base-pair transitions or transversions. Thus, it appears as though *M26* is a mutation that has created a recombinogenic sequence. Studies on *M26* with flanking markers should clarify what that mutation may tell us about mechanisms of recombination.

COMPARISONS OF MAPS WITH CHROMOSOMES

The existence of elements like *cog* and Chi implies that recombination probability need not be uniform along a chromosome. I would like to restate this for emphasis: map distances can relate to physical distances by a different factor in different regions of a chromosome. Several methods for comparing linkage map distances with physical distances have been developed for λ and T4. We encountered one of these earlier.

The density distribution of recombinants in replication-blocked crosses (Chapter 4 and Figure 9-3) compares physical with genetic distances at a glance. Of course, this method works only in the study of recombination systems, like Rec and Int, which work in the absence of chromosome replication. A rather small number of exchanges per chromosome must also obtain.

The normal linkage map of λ, obtained in standard crosses with Rec, Red, and Int functioning (Figure 2-7) has been compared with the chromosome by a different method. Viable deletion phages or phages carrying substitutions of heterologous DNA are the raw material for these studies. The end points on the linkage map of each of these chromosome aberrations can be located by recombination studies. Their end points on the chromosome can be determined by electron-microscopic analysis of heteroduplexes made in vitro between wild-type and aberrant DNA. The linkage map and the chromosome turn out to correspond pretty well,[58] indicating that recombination rates are rather uniform along the chromosome. (This conclusion ignores the obvious lack of correspondence in the neighborhood of *att*.)

In T4, *petite* particles with small capsids provide the raw material for an ingenious method of comparing the chromosome with the linkage map.[59] Since the amount of DNA packed in a T4 head is determined by the size of the capsid, *petite* heads have a constant, subnormal amount of DNA. Because the packaging process in T4 is insensitive to base sequence, a population of *petites* will contain all arcs of the cyclically permuted chromosome in equal amounts. The gene content of any *petite* particle can be assessed from the marker contributions it makes in co-infection with a normal particle. When the ends of the *petites* so dietermined are plotted along the linkage map, some regions of the map are found to be less densely represented than others. These under-represented regions must be those where the linkage map is elongated by a high rate of genetic recombination.

Transformation of T4-infected cells with genetically marked fragments of measured molecular weight provides analogous data for short distances.[60]

A method that is widely applicable in principle is comparison of numbers of amino acid residues between mutant sites in a protein with recombinant frequencies between the corresponding markers. For one genetically stretched region of T4, this approach has confirmed and extended the conclusions reached with *petite* particles.[61] Direct determination of base sequence in DNA by ever-improving methods promises to make other techniques for comparing map and chromosome obsolete. Furthermore, base sequencing will simultaneously provide more information on what makes the recombination rate variable along the chromosome.

Physical studies of fungi chromosomes are just beginning, so that the linkage map remains, with rare exceptions, our sole guide to the order of genes and the distances between them.

RECOMBINATION-LIKE REACTIONS IN VITRO

Several experiments support the plausibility of certain reaction sequences for recombination in vivo by demonstrating their reality in vitro.

Red exonuclease can incorporate a single chain of DNA into the structure of a duplex (chain assimilation). A substrate like that shown below was prepared and then treated with Red exonuclease:[62]

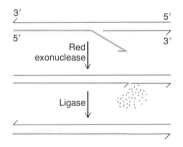

The exonuclease has digested the internal 5′ end until all superfluous DNA in the substrate is gone. An especially provocative feature of this reaction is that the exonuclease knows when to quit. As soon as the duplex has been trimmed, the enzyme dissociates from the duplex, thus avoiding the "undesirable" consequence of producing a gap in the duplex if it continued. By this in vitro demonstration, Red exonuclease becomes a good candidate for a late step in splatch formation.[63]

An early step in splatch formation may involve the "uptake" of homologous single-chain DNA by supercoils—a reaction we described earlier. The plausibility of that step is reinforced by the in vitro demonstration that the product of uptake can serve as a precursor to a recombinant duplex, like this:[64]

The in vitro uptake is catalyzed by purified RecA protein.[65]

An in vivo significance for the reaction is supported by the observation that DNA fragments of one genotype of phage φX174 plus supercoils of another yield recombinants when they transfect cells at about the same time. This recombination is appreciably $recA^+$ dependent.[66]

The application of one reaction for producing recombinants in vitro is revolutionizing genetic studies. In bacteria, a variety of endonucleases can chop DNA at sites having certain symmetric sequences. For example, the endonuclease EcoRI cuts at the sequence $\dfrac{GAATTC}{CTTAAG}$, yielding the products

At sufficiently low temperature the complementary single chains will anneal. If the annealed fragments are then ligated, the original molecule will be restored. Such splice formation by annealing is obviously recombinational.

A moment's reflection (and reference to Problem 5-3) reveals that this reaction has little relation to ordinary recombination. The complementary base sequence acted on by enzymes like EcoRI are so short that they occur repeatedly in all but the shortest viral chromosomes. The annealing reaction between the resulting fragments goes on without regard to the sequences on either side. Thus, the reaction yields inversions and translocations in vitro. While this feature of the reaction essentially eliminates it as a candidate for any role in recombination in nature, it is the reason why the reaction is used in so-called "recombinant DNA research." This ill-named activity involves the in vitro splicing of DNA without regard to phylogenetic relatedness and has produced viruses and plasmids into which genes from a wide variety of plants, animals, and microorganisms have been spliced. The usefulness of such bastards in research and technology has been and will be documented in other publications. (We have used the in vitro splicing of *E. coli* DNA fragments into λ to aid in our survey of the *E. coli* chromosome for Chi.[67])

PROBLEM

Propose a chemical model for genetic recombination. Make it as generally applicable as you can; at the same time be economical in your assumptions. List the phenomena explained by your model. Make another list of those which your model does not explain.

NOTES

1. B. McConnell and P. H. von Hippel (1970) *JMB* **50**:297.

2. R. B. Inman (1967) *JMB* **28**:103.

3. P. Doty, J. Marmur, J. Eigner, and C. Schildkraudt (1960) *PNAS* **46**:461; R. W. Davis and N. Davidson (1968) *PNAS* **60**:243.

4. C. S. Lee, R. W. Davis, and N. Davidson (1970) *JMB* **48**:1; T. R. Broker and I. R. Lehman (1971) *JMB* **60**:131; J. S. Kim, P. A. Sharp, and N. Davidson (1972) *PNAS* **69**:1948.

5. J. A. Berrman and J. Lebowitz (1973) *JMB* **79**:451.

6. W. K. Holloman, R. Wiegand, C. Hoessli, and C. M. Radding (1975) *PNAS* **72**:2394.

7. K. Mizuuchi and H. A. Nash (1976) *PNAS* **73**:3524.

8. M. Gellert, K. Mizuuchi, M. H. O'Dea, and H. A. Nash (1976) *PNAS* **73**:3872.

9. B. M. Alberts and L. Frey (1970) *N* **227**:1313.

10. A. Kornberg (1974) *DNA Synthesis*, San Francisco: W. H. Freeman and Company.

11. Y. Masamune and C. C. Richardson (1971) *JBC* **246**:2692; R. B. Kelly, N. R. Cozzarelli, M. P. Deutscher, I. R. Lehman, and A. Kornberg (1970) *JBC* 245:39.

12. M. P. Deutscher and A. Kornberg (1969) *JBC* **244**:3029.

13. M. Gellert (1967) *PNAS* **57**:148; A. Kornberg (1974) *DNA Synthesis*, San Francisco: W. H. Freeman and Company.

14. A. J. Clark and A. D. Margulies (1965) *PNAS* **53**:451.

15. G. Buttin and M. R. Wright (1968) *CSHSOB* **33**:259; M. Oishi (1968) *PNAS* **64**:1292; S. D. Barbour and A. J. Clark (1970) *PNAS* **65**:955; M. Wright G. Buttin, and J. Hurwitz (1971) *JBC* **246**:6543; P. J. Goldmark and S. Linn (1972) *JBC* **247**:1849.

16. J. Roberts, personal communication; L. Gudas and A. Pardee (1975) *PNAS* **72**:2330; K. McEntee (1977) *PNAS* **74**:5275; L. Gudas and D. Mount (1977) *PNAS* **74**:5280.

17. M. Abdel-Monem, H-F. Lauppe, J. Kartenbeck, H. Durwald, and H. Hoffman-Berling (1977) *JMB* **110**:667.

18. V. McKay and S. Linn (1976) *JBC* **251**:3716.

19. D. C. Eichler and I. R. Lehman (1977) *JBC* **252**:499.

20. A. E. Karu, Y. Sakaki H. Echols, and S. Linn (1975) *JBC* **250**:7377.

21. A. J. Clark (1974) *G* **78**:259.

22. A. J. Clark and A. D. Margulies (1965) *PNAS* **53**:451; P. Howard-Flanders and L. Theriot (1966) *G* **53**:1137.

23. A. Miura and J. Tomizawa (1968) *MGG* **103**:1; E. M. Witkin (1969) *MR* **8**:9.

24. K. Brooks and A. J. Clark (1967) *JV* **1**:283; I. Hertman and S. E. Luria (1967) *JMB* **23**:117.

25. A. J. Clark (1973) *ARG* **7**:67.

26. K. Brooks and A. J. Clark (1967) *JV* **1**:283.

27. H. Echols and R. Gingery (1968) *JMB* **34**:239; E. Signer and J. Weil (1968) *JMB* **34**:261.

28. J. W. Little (1967) *JBC* **242**:679; C. M. Radding, (1966) *JMB* **18**, 235.

29. M. J. Shulman, L. M. Hallick, H. Echols, and E. R. Signer (1970) *JMB* **52**:501; C. M. Radding (1970) *JMB* **52**:491; E. R. Signer, H. Echols, J. Weil, C. Radding, M. Shulman, L. Moore, and K. Manly (1968) *CSHSQB* **33**:711.

30. A. J. Clark (1974) *G* **78**:259.

31. Ibid; J. R. Gillen (1974) PhD thesis, University of California, Berkeley.

32. T. R. Broker and A. H. Doermann (1975) *ARG* **9**:213; R. C. Miller, Jr. (1975) *ARM* **29**:355.

33. J. Hosoda (1976) *JMB* **106**:277.

34. E. B. Konrad and I. R. Lehman (1975) *PNAS* **72**:2150.

35. M. I. Mosevitsky (1976) *JMB* **100**:219; W. D. Rupp, C. E. Wilde, D. L. Reno, and P. Howard-Flanders (1971) *JMB* **61**:25.

36. D. Freifelder (1966) *RR* **6** (Suppl):80.

37. F. W. Stahl (1956) *V* **2**:206.

38. M. A. Resnick (1975) in *Molecular Mechanisms for Repair of DNA* P. C. Hanawalt and R. B. Setlow, eds., New York: Plenum, p. 549.

39. A. Ahmad, W. K. Holloman, and R. Holliday (1975) *N* **258**:54.

40. P. Nevers and H-C. Spatz (1975) *MGG* **139**:233.

41. B. Rydberg (1977) *MGG* **152**:19.

42. D. G. Catcheside and D. Corcoran (1973) *AJBS* **26**:1337.

43. T. Angel, B. Austin, and D. G. Catcheside (1970) *AJBS* **23**:1229.

44. D. G. Catcheside and T. Angel (1974) *AJBS* **27**:219.

45. D. G. Catcheside (1975) *AJBS* **28**:213.

46. L. W. Enquist and A. Skalka (1973) *JMB* **75**:185.

47. J. Zissler, E. Signer, and F. Schaefer (1971) in *The Bacteriophage Lambda*, A. D. Hershey, ed. Cold Spring Harbor Laboratory, New York, p. 455.

48. D. Henderson and J. Weil (1974) in *Mechanisms in Recombination* R. F. Grell, ed., New York: Plenum, p. 89.

49. Ibid.

50. F. W. Stahl, J. M. Crasemann, and M. M. Stahl (1975) *JMB* **94**:203.

51. R. E. Malone, D. K. Chattoraj, D. H. Faulds, M. M. Stahl, and F. W. Stahl (1978) *JMB* **121**:473.

52. S. T. Lam, M. M. Stahl, K. D. McMilin, and F. W. Stahl (1974) *G* **77**:425.

53. F. W. Stahl and M. M. Stahl (1975) *MGG* **140**:29.

54. M. Radman, personal communication.

55. J. Siegel (1974) *JMB* **88**:619.

56. J. R. Gillen (1974) PhD thesis, University of California, Berkeley. F. W. Stahl and M. M. Stahl (1977) *G* **86**:715.

57. H. Gutz (1971) *G* **69**:317. S. L. Goldman (1974) *MGG* **132**:347.

58. N. Davidson and W. Szybalski (1971) in *The Bacteriophage Lambda* A. D. Hershey, ed., Cold Spring Harbor Laboratory, p. 45.

59. G. Mosig (1968) *G* **59**:137.

60. E. B. Goldberg (1966) *PNAS* **56**:1457.

61. S. K. Beckendorf and J. H. Wilson (1972) *V* **50**:315.

62. E. Cassuto and C. M. Radding (1971) *NNB* **229**:13.

63. C. M. Radding (1973) *ARG* **7**:87.

64. R. C. Wiegand, K. L. Beattie, W. K. Holloman, and C. M. Radding (1977) *JMB* **116**:805.

65. T. Shibata, C. DasGupta, R. Cunningham, and C. Radding, personal communication.

66. W. K. Holloman and C. M. Radding (1976) *PNAS* **73**:3910.

66. W. K. Holloman and C. M. Radding (1976) *PNAS* **73**:3910.

67. R. E. Malone, D. K. Chattoraj, D. H. Faulds, M. M. Stahl, and F. W. Stahl (1978) *JMB* **121**:473.

10

Models for
Phage Recombination

The relative ease with which phage chromosomes can be studied by physical and chemical as well as genetic methods has made possible relatively detailed models for recombination. Recombination systems differ sufficiently to lead us to propose quite a variety of schemes. These schemes have little in common beyond the breaking and joining of DNA, with or without local or extensive DNA synthesis. They provide us with a variety of possibilities to play with when we try to model fungal recombination in the next chapter.

RULES FOR MODEL-MAKING

Two general rules apply to all models. (1) A model should fit the data. It need not fit all the data, since the chances are that some of the data are artifactual. Nevertheless, it should fit the well-established facts and enough of the rest to hold promise of wide applicability. (2) A model should rest on a foundation of widely applicable central assumptions. A model's usefulness will be judged by the degree to which it helps us define that which is common to all genetic recombination in all creatures. We must be alert, however, to the possibility that this common core is a small one and be tolerant of alternate models when good experiments call for them.[1]

For genetic recombination models in particular, all assumptions must be consistent with the physical and chemical properties of DNA (or RNA, if we were seeking to explain recombination in those few RNA viruses that have shown it.) The model will be even more attractive if it makes use of demonstrated in vitro reactivities of DNA. Good models of recombination make maximum use of enzyme reactions demonstrated in vitro and minimal use of hypothetical reactions whose only qualifications are that they are thermodynamically sound, mechanistically conceivable, and are needed for the model.

The standards for recombination models are therefore of two sorts—the timeless standards set by the laws of chemistry and the ever-shifting ones erected by our changing knowledge of DNA biochemistry. Let me illustrate. Fifteen years ago a model that supposed a role in recombination for supercoiled DNA helices would not be admitted to polite circles. Though supercoils were obviously compatible with the restraints of thermodynamics and structural chemistry, they had not been demonstrated nor even discussed, so that a postulated role for them in recombination would have appeared embarrassingly ad hoc. Eventually, supercoils were demonstrated in isolated DNA, making a recombination model that featured them more plausible. The admissability of the model was debated, however, because of the possibility that supercoils did not exist in vivo. Now that an enzyme that introduces supercoiling has been demonstrated,[2] supercoiling will become absolutely *de rigueur* in models. In fact, a role for supercoiling in the initiation of generalized recombination has already been postulated[3] and partially demonstrated, and its role in site-specific recombination has been established.[4]

Most recombination models have been based on genetic data, i.e., on recombination and conversion frequencies. These models have been ingeniously devised to fit such data. For various reasons, especially the intrusion of short-distance marker effects, we are rarely able to distinguish among the models by genetic data alone. The models do, however, frequently make distinctive physical predictions, most notably with regard to the involvement of DNA synthesis in the recombination processes. Such physical predictions are at present not verifiable in eukaryotes; the essential technology is only on the horizon. With phages, however, some progress has been made and more should soon be forthcoming. For that reason we shall take up models for phage recombination first. In the next chapter we shall examine models for fungi, using the partially tested ones from phage to moderate our fancies.

ACTION OF THE REC SYSTEM ON LAMBDA

When λ strains defective in their *red* and *gam* (and *int*) genes are crossed, the resulting recombinants arise by the action of the Rec system of the *E. coli* host; i.e., the recombination is almost totally dependent upon $recA^+$ function[5] and appreciably (about 70 percent) dependent on the $recB^+$ and C^+ ones as well.[6] The recombination is more or less uniform along the λ chromosome and it occurs with no obvious dependence on chromosome replication. Uniformity is concluded directly from the approximately uniform density distribution of recombinants between terminal markers in density-labeled, replication-blocked crosses (Figure 10-1). The lack of

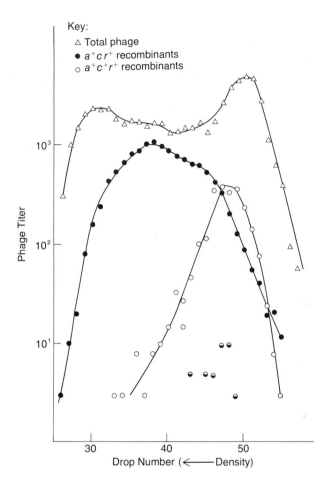

Key:
△ Total phage
● $a^+ c\, r^+$ recombinants
○ $a^+ c^+ r^+$ recombinants

FIGURE 10-1

The density distributions of λ produced in rec^+ red gam int crosses with DNA replication blocked. The cross was

The $a^+ r^+$ recombinants are approximately uniformly distributed from the fully heavy position near drop 30 to the light one at drop 50. Thus, with replication blocked, exchange is almost uniformly probable along the entire chromosome. (F. W. Stahl, K. D. McMilin, M. M. Stahl, J. M. Crasemann, and S. Lam [1974] *G* **77**: 395.)

189

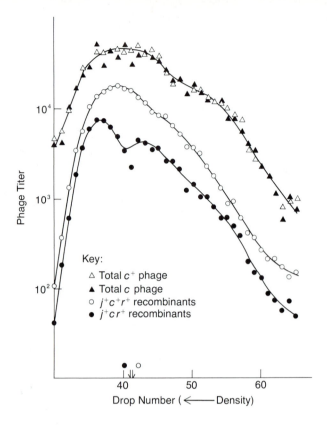

FIGURE 10-2
The density distributions of λ produced in a *rec⁺ red gam int*
cross allowing a little DNA replication. The cross was

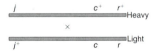

j^+r^+ recombinants arising by exchange near the middle of the
chromosome carry the c^+ marker; those arising near the right end
carry the c marker. The two classes of recombinants have the
same average density, indicating that they are not differentially
dependent on DNA replication. (F. W. Stahl, K. D. McMilin,
M. M. Stahl, J. M. Crasemann, and S. Lam [1974] *G* 77:395.)

dependence on replication is less directly supported. The argument is in two
parts. (1) As shown in Figure 10-2, a central region and a terminal region show
no differential response to replication. We shall see later that this result is
significantly in contrast to crosses involving Red, in which these two regions
differ strongly in their dependence on chromosome replication. (2) The Int
system makes a small but detectable contribution in both replication-

blocked and ordinary crosses.[7] The contribution is comparable in the two cases, indicating that Rec is no more stimulated by replication than is Int. In view of Int's dependence on a nonreplicating, supercoiled substrate (as described later in this chapter), it seems unlikely that a dependence of Rec upon a replicating one would allow the observed relationships. Thus, Rec is a break-join system in the classic sense (Figure 1-3).

In crosses involving an active *gam* gene, Rec (presumably the RecF pathway) has been shown to be almost invariably reciprocal in single bursts.[8] In other kinds of experiments the Rec system is frequently reciprocal when the *recBC⁺* nuclease is presumed active as well (i.e., when the RecBC pathway is operating.)[9]

Our attention is focused on one class of reciprocal break-join models by the demonstration of so-called figure eights.[11] In the presence of the *recA⁺* function, the circular *E. coli* episome colE1 gives rise to dimers with this structure:[12]

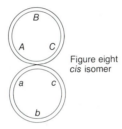

Figure eight
cis isomer

The detailed structure at the junction can be seen in molecules in which one circle is rotated 180° with respect to the other:

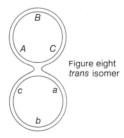

Figure eight
trans isomer

The rather open nature of the junction in the micrograph in Figure 10-3 was achieved by mild denaturing conditions. Three-dimensional physical models indicate that a junction in which all bases are hydrogen bonded to their complements can exist (Figure 10-4).[13]

FIGURE 10-3
Electron micrograph of a "Holliday structure" in the bacterial
episome colE1. This dimer with four ends was derived in vitro by
cutting a figure eight with an enzyme that chops colE1 just once
at a specific sequence. The DNA was spread for microscopy
under mild denaturing conditions, causing the junction to open
up a bit. Compare this photograph with the diagram of a figure
eight trans isomer on page 191. (H. Potter and D. Dressler
[1976] *PNAS* **73**: 3000.)

FIGURE 10-4 (*opposite*)
Basic features of the stereochemical interconversion of bridging chains in a Holliday structure.
   Top. In order to interconvert bridging chains in a Holliday structure, one must pass
through a fourfold junction.
   Bottom. In order to convert a Holliday structure to a fourfold junction, one rotates DNA
sections above and below the crossed-chain exchange in a counterclockwise manner, simultaneously bending them apart. The fourfold junction can then be converted to an identical
Holliday structure with new bridging chains by continuing the combined operations of
rotation and bending of DNA sections in the same sense. Before interconversion, section A
lies above C, and B above D. After interconversion, section D lies above A, and C above B;
DNA sections formerly on top of the crossed-chain exchange (i.e., A and B) lie below the
new, structurally identical, crossed-chain exchange. (H. M. Sobell [1974] in *Mechanisms in
Recombination*, R. F. Grell, ed., Plenum Publishing Corporation, New York, p. 433.)

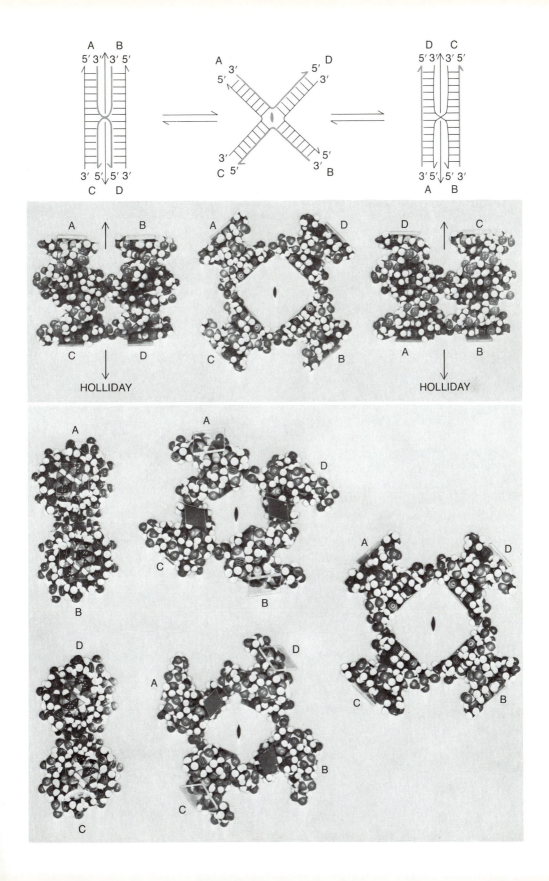

Figure eights have been seen in lambda[14] and in the little phage ϕ X174.[15] In ϕ X174 one monomer was demonstrably derived from a different infecting parent than the other.[16] Thus, we presume that these structures are intermediates in a pathway leading to recombinants.

Studies of figure eights[17] indicate that in vitro they are subject to branch migration, so that the dimeric chain is dynamically shared between the two monomeric ones:

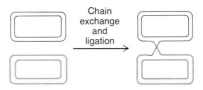

These data on Rec lead one to a model that is similar in its outlines to the hypothesis orginally propounded by Holliday.

By mechanisms we shall speculate upon later, single chains are presumed to undergo reciprocal exchange of partners and be joined by ligase (if cuts on the two circles are not at precisely homologous spots, then the overall reaction must involve "trimming" by a nuclease and "repair" by a polymerase):

The resulting splatches can then be elongated by branch migration in either direction:[18]

Resolution of a figure eight can occur in either of two ways, which appear to be equally probable, as we shall explain. I like to imagine that the first step in resolution is the migration of the junction or cross point to an "accidental" nick, i.e., a nick (or gap) that occurs in the figure eight. (Alternatively, the cross point could be especially nuclease-sensitive despite the model-building evidence (Figure 10-4) that it need contain no single-chain regions.) The nick can be on either the dimeric chain or a monomeric one. For this diagram, we draw the trans isomers:

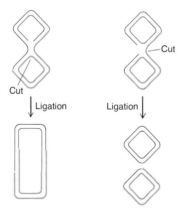

I imagine that the nick exposes the opposite chain in the Holliday junction to endonucleolytic cutting, perhaps by the $recBC^+$ enzyme. The products are a dimer circle containing two splices or two monomers, each with a patch:

For the somewhat special case of Rec on λ, the fates of the dimer and the monomer must be separately considered. A recombinant component of a dimer circle can be encapsidated directly from the circle. A monomeric recombinant, however, must become part of a larger structure to be encapsidated. Such a monomer, like any other, can become eligible for encapsidation in one of two ways; either it can replicate in the rolling-circle mode, or it can splice into another λ circle. In a *red gam* infection (in rec^+ cells), rolling-circle replication appears to be blocked, and we must presume that all recombinants become encapsidatable by the splice-formation reaction. Any experimental effort to assess the relative frequency of splices and patches in λ is likely to be obfuscated by this requirement.

The model as presented above is silent about molecular mechanisms promoting chain exchange, i.e., about the early steps in splatch formation. An attractive possibility for initiation of λ recombination by the Rec system is suggested by the Rec-dependent recombination between supercoils and DNA fragments introduced into cells along with them (Chapter 9). In the diagrams below, a nick or small gap is acted upon by either of two enzymes, with the common consequence that a chain is displaced from a duplex. A chain can be displaced by a polymerase polymerizing nucleotides onto a 3'OH chain end or by the $recBC^+$ nuclease in the presence of DNA-annealing protein (RecA?). The displaced chain is then taken up by a supercoiled circle:

Cut by
endonuclease

The stretch of polynucleotide in the supercoil that has lost its pairing partner is supposed to be endonuclease ($recBC^+$ nuclease?) sensitive as indicated in the above diagram. Following the cut, annealing between the two single chains produces the figure eight:

and ligation may close it covalently.

Chi in λ is a good substrate for the Rec system, increasing the rate of exchange in its neighborhood. Its activity requires $recA^+$ and $recBC^+$ function. The role of $recA^+$ remains mysterious, but we can venture some thoughts on the possible role(s) of the $recBC^+$ nuclease in Chi activity.

The *recBC*$^+$ nuclease might recognize the Chi sequence on a DNA duplex and make a single-chain nick. The ability of *recBC*$^+$ nuclease to act endo-nucleolytically on double-chain DNA has not been demonstrated, so this proposal is somewhat out of order. On the other hand, sequence-specific endonucleolytic activity by *recBC*$^+$ nuclease might well have been over-looked, and proposing ExoV for the role of Chi-cutter at least relieves me of the embarrassment of having to pull an enzyme out of a hat. Whether or not the *recBC*$^+$ nuclease makes the presumed initial cut at Chi, it does seem like a thoroughly natural choice for catalyzing the next step. Somehow, Chi can effect exchange at a distance. It can do so when the other parent chromo-some is deleted or heterologously substituted for the Chi region. Apparently, Chi initiates an event that travels along the chromosome and entrains the second chromosome at some remove from Chi. Gapping, or perhaps chain-displacement, by ExoV could be the traveling event. Alternatively, of course, chain extension by DNA polymerase could serve, by either nick translation or chain displacement.

Within this generalized proposal for the role of the *recBC*$^+$ enzyme in Chi activity, there are two possibilities for the time of Chi's action: Chi might promote the initiation of figure eights or it might promote the conver-sion of figure eights to recombinant dimers and (perhaps) monomers.

If Chi acts to initiate figure eight formation, there are two ways in which it might do so. The observation that Chi is highly active even when oppo-site a deletion tells us that the putative initiation reaction of Chi must be asymmetric—Chi does something and the other duplex responds. The "something" that Chi does, it seems to me, could be either "passive" or "aggressive." If a gap is initiated at Chi, then that gap might act as an acceptor of single-chain feelers sent out (independently of Chi) by other duplexes.

(Subsequent steps in splatch formation could then proceed in any of several ways). In this scheme Chi initiates a kind of FPS within which multiple exchanges could occur.

Chi might act "aggressively," on the other hand, by itself providing free ends. For instance, the free ends at putative Chi-initiated gaps are apt to be melted out (or driven out by the ExoV displacement reaction) and available for uptake by supercoiled recipients:

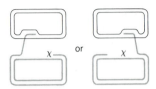

Subsequent steps might then proceed as we described earlier for Rec-mediated recombination. In this scheme Chi is acting as a fixed point for splatch formation while generating a variable point in its neighborhood at the same time. (If gapping proceeded in both directions from Chi, both ends of the gap would be variable and neither splatch would be truly fixed with regard to its starting point.)

The other possible role for Chi in recombination is that it acts to resolve figure eights, by stimulating formation of a translating nick or a gap that leads to resolution of the figure eights.

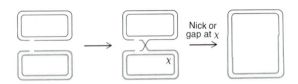

There is experimental support for the possibility that the $recBC^+$ nuclease, prime suspect for a direct role in Chi activity, acts late in recombination, after heteroduplex formation. During bacterial conjugation, recombination between mutant markers in the same gene can be measured by the appearance of a wild-type enzyme. Although $recBC$ bacterial crosses give an almost normal recombination rate as measured by enzyme production, they produce few viable recombinant progeny.[19] We may postulate that the recombination resulting in a wild-type gene occurs by mismatch correction within un-resolved splatches. In the absence of $recBC$ nuclease, most of the splatches never do resolve to splices or patches, and the recombinants die.

REC ON LAMBDA IN THE ABSENCE OF $recBC^+$ NUCLEASE

When ExoV is inactivated by either mutation or *gam* product, recombination becomes regionally dependent upon replication. In crosses fully blocked for replication (Figure 4-3), more recombinants are found near the chromosome tips than are seen in *red gam* crosses (Figure 10-1). When the block on repli-

Models for Phage Recombination

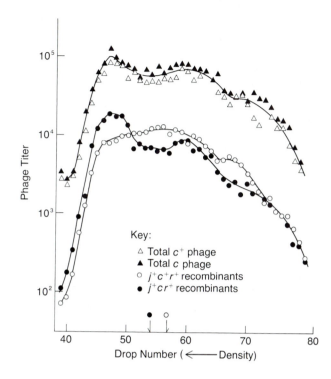

FIGURE 10-5
The density distributions of λ produced in a *rec⁺ red gam⁺ int* cross allowing a little DNA replication. Markers are the same as in Figure 10-2. The $j^+c^+r^+$ recombinants have a lower average density than the j^+cr^+ ones, indicating that recombinants arising by exchange near the middle of the chromosomes have enjoyed more DNA synthesis than those arising near the right end. (F. W. Stahl, K. D. McMilin, M. M. Stahl, J. M. Crasemann, and S. Lam [1974] *G* 77:395.)

Key:
△ Total c^+ phage
▲ Total c phage
○ $j^+c^+r^+$ recombinants
● j^+cr^+ recombinants

cation is partly lifted, recombinant formation in the middle of the chromosome catches up with that at the ends (Figure 10-5). We may understand these phenomena by supposing that *recBC⁺* nuclease does indeed play a role in the resolution of figure eights, as shown in this diagram:

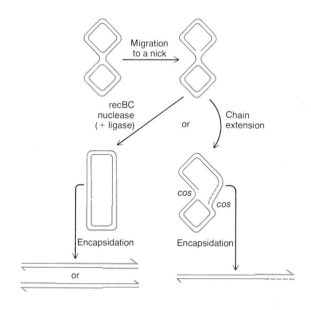

A duplex that lies between two *cos* sites can be encapsidated. When ExoV is active, a major route to the production of such packageable chromosomes is by simple endonucleolytic resolution of figure eights to recombinant dimer circles. This reaction may involve little or no DNA synthesis. When ExoV is inactive, the rate at which figure eights are resolved to dimeric circles is reduced. In this situation polymerase can add nucleotides to the 3′ end of the nicked circle. But in the absence of a full replication apparatus, chain extension can only proceed a limited distance (perhaps a few thousand bases). If it reaches a *cos* site, a chromosome can be encapsidated. This chromosome is most apt to be recombinant near an end, as shown in the above diagram. (Implicit in this contention is the assumption that accidental nicks occur at a high enough frequency that most splatches do not get very long.) Thus, in the absence of *recBC*$^+$ nuclease, recombinants arising near the chromosome termini are favored when replication is blocked. What if replication is permitted? Then chain extension at the 3′ end of the nicked circle graduates into a full-fledged replication fork, like this:

When the replication fork reaches a *cos* site, the recombinant can be encapsidated. Thus, encapsidated chromosomes that are recombinant for a central interval will be of intermediate density in a density-label transfer cross, reflecting extensive DNA synthesis. Their structure is "break-join" on one chain and "break-copy" on the other (Figure 1-3).

ALL WILD-TYPE RECOMBINATION GENES ACTING ON LAMBDA

When wild-type λ infects wild-type *E. coli*, the Red system works in conjunction with Rec whose *recBC*$^+$ nuclease is *gam*-inactivated.[20] In replication-blocked crosses recombination at the chromosome ends is considerably

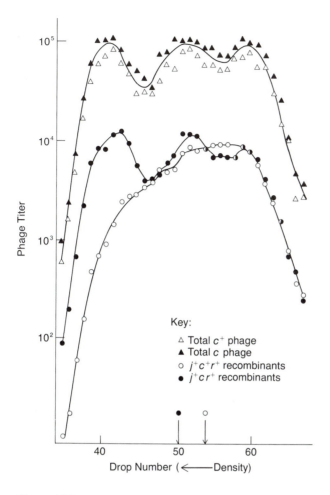

FIGURE 10-6
The density distributions of λ produced in a *rec*⁺ *red*⁺ *gam*⁺ *int*
cross allowing a little DNA replication. Markers are the same as
in Figure 10-5. The $j^+c^+r^+$ recombinants have a much lower
average density, indicating that recombinants arising near the
middle of the chromosome have enjoyed much more DNA
synthesis than ones arising near the right end. (F. W. Stahl,
K. D. McMilin, M. M. Stahl, J. M. Crasemann, and S. Lam
[1974] *G* **77**:395.)

more frequent than recombination in central regions (Figure 4-5), and that
in central regions is considerably replication dependent (Figure 10-6). Red
thus accentuates the phenomenon seen in *red gam*⁺ crosses, and economy
prompts us to seek an explanation within the same framework.

We suppose that chain extension is facilitated by the strategic removal of DNA, catalyzed by the Red 5′ exonuclease:

A *cos* that would have been unreachable by chain extension alone becomes reachable with the aid of Red exonuclease. The single chain exposed by exonuclease action then pairs with its monomeric complement, permitting chain extension to proceed more effectively. As specified above, the model appears capable of generating recombinants anywhere in the chromosome without the intervention of replication forks. Indeed, it may do so, but the probability of success for a splice far from the chromosome end is reduced by the possibility that the Red exonuclease progression is interrupted by an accidental nick. We suppose that, as the Red exonuclease approaches a nick, the short duplex region ahead of the enzyme melts out, leaving the enzyme holding a single-chain tail. Red exonuclease's low affinity for single-chain DNA results in dissociation of the enzyme from the dimer:

Note that this structure has the topology of a rolling circle (though its chain polarities are unconventional.)[21] This feature of the model harmonizes with the observation that Red promotes late, rolling-circle replication. The Rec system also promotes replication (at least in *gam*[+] or *recBC* infections) by initiating replication forks (see the diagrams on p. 199 and p. 200).

In Chapter 6 we discussed experiments that bear on this model for the joint activity of Rec and Red. In replication-blocked crosses phages hetero-duplex at genes *O* and/or *P* and having a minor material contribution from one parent (say the heavy one) were found to be primarily of this sort

rather than this

The maturable recombinant chromosome in the diagram at the top of page 202 is like the chromosome in the first diagram above because I rigged it to be so. What if instead I had drawn the earlier diagram like this?

Now the maturable recombinant chromosome would have a structure like that in the second diagram above. In the model as defined, there is no reason why the alternate configurations should occur at unequal frequencies.[22] What force can we call upon that might account for the observed bias in the data?

One possibly relevant asymmetry in λ chromosome behavior has been noted—chromosome encapsidation begins at the *cos* near gene *A*, at the left end of the chromosome. Since recombination is likely to be contemporaneous with encapsidation, the recombinational intermediate in the diagram on page 202 may well be in the process of being encapsidated near the *A* end of the ultimate recombinant product while it is being completed by replication at the *R* end. If such were the case, the events depicted in the diagram on this page could be depressed, and the preponderance of heteroduplexes near the right end would be those in the diagram on page 202 (top).

RED ACTING ALONE ON LAMBDA

In ordinary crosses in *recA* cells the Red system yields recombinant frequencies almost as great as those observed in *recA*[+] *red*[+] complexes—Red shows little or no dependence on *recA*[+] function. In replication-blocked crosses, however, a different situation is observed. In a replication-blocked cross recombination is the only route to dimer formation and encapsidation. Red acting in *recA* cells produces far fewer mature phages than it does in the presence of *recA*[+] function.[23] In fact, Red alone produces fewer particles in a replication-blocked cross than does Rec alone. The low yield in *red*[+] *recA* replication-blocked crosses is not due to a destructive activity of ExoV; *red*[+] *recA recBC* crosses give as low a yield of phage as do the *recBC*[+] ones.[24] Nor is it due to a failure of the *red*[+] genes to make gene product; the ability of the replication-blocked phages to express Red function is evident from the appreciable contribution the Red system makes in a *rec*[+] replication-blocked cross. Thus, it appears as though Red has some requirement that can be met by either replication or *recA*[+] function. Perhaps that requirement is the exposure of single chains.

The few recombinants that are produced by Red alone in replication-blocked crosses occur near the right end of the λ chromosome and have a splice polarity like that seen in the *red*[+] *rec*[+] crosses.[25] Furthermore, when a little replication is allowed, centrally located intervals are seen to respond relatively more than the right end.[26] These results would be explained if Red required either *recA*[+] or replication to produce figure eights.

The possibility that *recA*[+] and replication can substitute for each other in Red recombination is suggested by another line of evidence. As described in Chapter 9, the recombination pathway RecE, activated by the *sbcA* mutation, appears to be homologous to Red. The RecE pathway recombines λ *red* chromosomes infecting *recA* cells. During bacterial conjugation, however, *E. coli* recombination by the RecE pathway is absolutely dependent on *recA*[+] function. Why this difference? An economical supposition is that bacterial chromosomes do not replicate during conjugational recombination, whereas λ chromosomes do replicate during a phage cross. This view provides one of the few (flimsy) grounds for speculating on the role of *recA*[+] product; it may act early in recombination, doing to a chromosome *something* that can also be done by replication.

Conspicuous features of the Rec and Red systems are the reciprocality of the former and the apparent nonreciprocality of the latter. Do the models presented account for these features? For the case of Rec, the symmetry of the figure eight is a good start toward reciprocality. If resolution of the figure eight is also symmetric, then some degree of reciprocality is assured. We must consider three modes of resolution.

1. Resolution of a figure eight to two monomeric circles each with a patch will give circular monomers whose degree of reciprocality will reflect the symmetry of the process that formed the figure eight. Some of the reciprocality will fail to be manifested among mature progeny, however; different mismatch corrections or different degrees of luck at replication and maturation of the two monomers will tend to obscure the reciprocality.

2. If the figure eight resolves to a dimer circle, the manifestation of reciprocality among mature progeny will depend upon replication of the dimer. From any individual dimer, only one or the other moiety can be encapsidated.[27] The λ that is left behind can be matured only if it, or its descendants, become part of a multimer by rolling-circle replication or exchange. Reciprocality can be obscured in either direction: the monomer left behind may fail to leave any representatives among mature particles or, conversely, the monomer left behind may exploit the opportunity to multiply and succeed in leaving mature progeny that greatly outnumber the one copy of the complementary recombinant matured from the initial dimer. If, on the other hand, dimers generally replicate several times *before* being subject to maturation, the two moieties will usually be comparably represented among progeny particles and the reciprocality of the exchange act will be evident.

3. If the Holliday junction becomes a replicating fork, it appears that the reciprocality can be preserved. The product of full replication will be a recombinant dimer plus a parental monomer:

The recombinant dimer will manifest its reciprocality among mature progeny in accord with the conditions discussed above.

Our model for Red (as well as for Red-plus-Rec) likewise proposes a figure eight intermediate. However, one loop of the eight is partially destroyed by 5′ exonuclease, perhaps rendering the overall process nonreciprocal.

Despite the documented differences between Red and Rec behavior, noteworthy similarities exist between the two systems. First, "for both systems [Red and *gam*-depressed Rec, presumably RecF] the average clone size of double recombinants is [about] half that of singles."[28] This result not only indicates a common aspect of the two systems but bears as well on the mating theory described in Chapter 2. The λ linkage map can be rectified with a mating theory mapping function that supposes a rather small number of matings per progeny particle. This is tantamount to saying that most double recombinants arise in a single mating act. In that case, the mating theory leads us to expect the same average size for clones of double recombinants as for single recombinants, since they arise by the same kinetics. The observed twofold difference in clone size reveals a paradox. I think the solution to the paradox is likely to lie in one of the following two considerations. (1) Single-exchange products may replicate faster than double-exchange products. Double exchange between two circles (whether by patch formation or by paired splices) yields two monomer circles as products; single exchange, on the other hand, yields a dimer. Dimers are apt to differ from monomers with respect to replication (they have *two* replication origins!) and certainly differ with respect to maturation (they are concatamers). (2) A λ mating may take longer than we supposed; in particular, those with two exchanges may take longer to complete than those with one, so that double recombinants emerge late from the "mating room." A special case of such an explanation is again to be found in the circularity of lambda's chromosome. The onset of a "mating" may be an exchange that unites two monomers to give a dimer. This dimer may replicate. Before long, a daughter dimer is very likely to resolve back to monomers by a second exchange *facilitated* by the proximity of the two members of the dimer. This facilitation accounts for the exchange correlation (negative interference) that necessitated the mating theory. Since a replication sometimes intervenes between the onset and completion of a "mating," double recombinants will have smaller clone sizes than singles. Until these possibilities are resolved and quantitated, the beguiling relations between the clone sizes cannot be interpreted in terms of the mechanism of exchange per se.

"Another similarity we find for Rec- and Red-mediated recombination has to do with the local distribution of exchanges on the phage chromosome . . . . If a phage is recombinant for one interval, its likelihood of also being recombinant in the adjacent interval is the same for the two recombination systems, even though the overall frequency of recombination is three times greater for Red than for Rec. It is as though the parameters of the recombination mechanism that govern the relative frequency of single and double recombinants are the same for the two systems."[29] Such a result has a simple formal explanation within the framework of the mating theory, namely, that the higher activity of Red is due solely to Red's encouragement of a higher number of matings per particle. If we abandon the mating theory, however, we are free to seek an explanation in the models for Red and Rec discussed in this chapter. It was suggested that both Red and Rec produce figure eights. The double crossovers seen in λ may be due mostly to eights that resolve to give patches (more or less reciprocally, at least in the case of Rec). If the ratio of splices to patches and the length distribution of splatches is the same for Rec and Red, then the relationship quoted will hold. Termination of splatches by accidental nicks might ensure the required conditions.

RECOMBINATION BY INT

Int-mediated exchange in λ occurs efficiently in vitro under appropriate circumstances. A high local concentration of sites (*att*) of Int action is ensured by the use of λ chromosomes containing two *att* sites. There is sufficient DNA between the sites to permit bending of the chromosome to bring them into apposition:

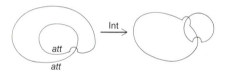

The reaction requires ATP and occurs only when the chromosome is in a nickfree circular state.[30] These two requirements are necessary for the chromosome to be put into an underwound, supercoiled state essential for the reaction. The underwinding, i.e., fewer twists per unit length than called for by the Watson-Crick structure, is catalyzed by an enzyme (DNA gyrase) and driven by the splitting of ATP. The underwinding *probably* facilitates exchange of chains in a special short region in *att*. Most of the underwinding is retained in the products of the reaction, which consequently remain both interlocked and supercoiled in this in vitro imitation of the in vivo Int-reaction. (In vivo, the products are invariably, and mysteriously, resolved!) The retention of supercoiling indicates that the Int enzymes hold on to the two *att* sites throughout the exchange of chains and the reformation of all phosphate-sugar bonds. One can imagine that the reaction occurs as shown in Figure 10-7.

Determination of nucleotide sequences shows that *att* on λ has a 15 base-pair sequence identical to one in the *att* site on *E. coli*.[31] Furthermore, ingenious genetic studies support the idea that short splices are in fact involved in Int-mediated exchange. The first indication of splices in Int-mediated exchange came from studies of mutant *att* sites.[32] Int-mediated

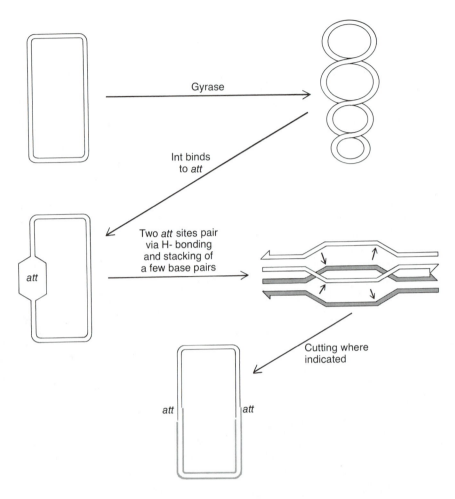

FIGURE 10-7
A speculative summary of Int-mediated recombination.

exchange occurring between mutant and wild-type *att* gave progeny particles that were heterozygous. The second indication came from crosses involving markers less than 1,000 base pairs away from *att*.[33] The crosses were like this:

h	*att int+*	*iλ*

×

h+	*att int*	*i434*

Both parents were *red* and the crosses were performed in *recA* bacteria. A mutation in *int* was used as the marker, but recessivity of the mutation assured the functioning of the Int system. The hi^{434} recombinant was selected, and, as expected, most of the recombinants carried the *int* marker; they appear to have exchanged at the *att* site. However, a small fraction of the hi^{434} recombinants were int^+.

Your first guess might be that these recombinants arose because of "leaky" genetic blocks in the Red or Rec pathways. Not so. When Int function was blocked by appropriate mutations, these recombinants disappeared. Furthermore, the number of these recombinants fell rapidly as markers further away from *att* were used. Markers in the *xis* gene, just to the right of *int*, gave a barely detectable frequency of recombinants in the interval to their right. Apparently, the Int-initiated exchange can sometimes wander outside of its usual domain, the *att* site, and effect recombination there. The leading candidate for the mechanism of wandering has to be elongation of reciprocal splatches by branch migration. By this view, *att* is a fixed starting point for splatch formation. Unknown special features of the system usually confine exchange to that fixed point, but not always.

RECOMBINATION IN T4

A bright idea provides a unifying framework for the roles of the T4 terminal redundancy in chromosome replication and the roles of T4 chromosome ends in genetic recombination.[34] The conventional picture of chromosome replication diagrams a replicating fork like this:

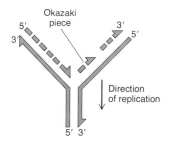

The new chains are synthesized by adding nucleotides to the 3′OH ends of the growing chains. This restriction implies that one new chain (on the left) can grow continuously, toward the fork, while the other new chain grows away from the fork and must be repeatedly initiated as the fork progresses.

For a noncircular but permuted chromosome, like that of T4, ends would appear to pose a special problem.

If initiation of Okazaki pieces required special sequences, then most of the T4 chromosomes would be unable to replicate their entire length. When a fork reached the end of a chromosome, the products would look like this:

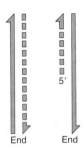

In fact, replication in T4 begins internally in each chromosome (at several places) and proceeds to the ends, so that both daughter chromosomes must end up with single-chain ends, like this:

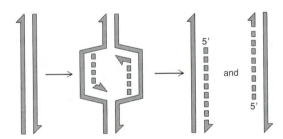

Repeated rounds of replication, in this way, would lead to ever shorter chromosomes. Something must be done about that! What could be better than head-to-tail annealing between complementary chains in their terminal redundancies, like this:

This reaction gives a molecule that is approximately dimeric. The loss of terminal information in this process accounts for the polarized segregation of redundancy heterozygotes; the complementary redundant single chains provide the basis for the high rate of head-to-tail genetic recombination among daughter chromosomes (Chapter 4). The terminal single chains also provide a starting point for a general recombination model for T4, in which, as you recall, most recombination is focused at the ever-permuting chromosome ends. The exposed end of a chromosome may be taken up by a negatively supercoiled homologous stretch of T4, in the fashion we considered for Rec acting on λ:

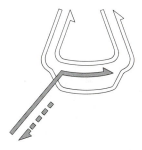

The displaced chain could be removed enzymatically, giving structures like those microscopically identified in T4-infected cells (Chapter 4):*

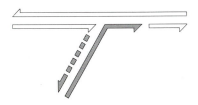

Resolution by endonuclease cutting can be to either a patch or a splice. Either way, the asymmetry of the structure assures nonreciprocality in the production of complementary recombinant types.

At first glance, the supercoil proposal may appear farfetched for T4, since supercoiling is a property we tend to associate with circular chromosomes. However, the possibility cannot be discounted that segments of DNA physically constrained, perhaps by RNA, supercoil in a fashion that has been demonstrated for the RNA-constrained loops of *E. coli*.[35] However,

---

* In the electronmicrographs of replication-blocked T4 in Chapter 4, the single-chained ends cannot have arisen as replication by-products; they must be the result of exonuclease activity.

any involvement of such supercoiled loops in T4 recombination is speculative. It may be that single-chain ends pair with single chains exposed by internal gaps in another chromosome. Support for this proposal comes from the observation that conditions that increase the length and lifetime of gaps are recombinogenic in T4. In most creatures the recombinogenic role of ultraviolet-induced damage, whose repair leads to gaps, argues for a role for gaps in genetic recombination.

One source of LNI in T4 falls quite simply out of this model. Since a T4-infected cell will contain a number of sister chromosomes with ends in about the same place, any chromosome is subject to "attack" by a number of ends in the same region. We do have to consider at least one alternative explanation for LNI, however.

In Chapter 4 we discussed electron-micrographs that implied LNI. These pictures of branched recombinational intermediates show a nonrandom distribution of branches. In fact, the branches are clustered in a fashion reminiscent of the clustered recombination events we call LNI. Since these molecules were from replication-blocked infections, the explanation for LNI offered above cannot apply to this clustering of branches. Another kind of explanation has been offered. Some of the branched T4 DNA structures differ from the one drawn above. These molecules appear to have arisen by annealing at internal gaps on each of two molecules:

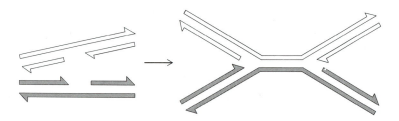

The two participants, held together by a common stretch of Watson-Crick duplex, may now gap and anneal with each other in a nearby region in a reaction facilitated by mere proximity:

Endonucleolytic cutting that resolves the structure can leave us with a multiply exchanged chromosome. However, I think (with support[36]) that normal LNI is due mostly to the presence of sister chromosomes with similar ends in individual cells.

SOME OTHER CONSIDERATIONS

Under certain circumstances exchange is especially vital for the production of virions in both T4 and λ. When T4 crosses are performed between *petite* particles, the production of full-size chromosomes requires exchange between two short ones, presumably near the end of one of them. In order for the exchange to be effective in promoting the formation of packageable DNA, it must result in a splice, not a patch, and must involve a single rather than a double exchange. Crosses between *petite* chromosomes might thus be expected to show interference relationships different from those in ordinary crosses. Indeed, positive interference characterizes such crosses.[37] This positive interference must result in part from a reduction in "mating heterogeneity" imposed by the requirement of recombination; no incomplete chromosome that fails to exchange with a chromosome from a different clone can contribute to viable progeny.

Abnormal interference relations in λ can be anticipated in replication-blocked crosses. Indeed, terminal markers give close to 50 percent recombinants (Figure 4-3) even though the number of exchanges per chromosome is rarely greater than one. In *red gam* crosses, our picture of the λ life cycle likewise requires that each virion chromosome come from a mating involving a single crossover (taking triples, etc., to be rare). Again, about 50 percent recombinants is expected for terminal markers, but only about 20 percent is found. Two explanations seem plausible. (1) Some chromosomes may achieve dimeric status by a $recA^+$-dependent reaction that is replicational, not recombinational, in nature. (2) Intraclonal mating may be frequent in λ, at least under these conditions. (This assumption implies a small effective value for g in the mapping function. See Problems 4-6 and 10-3.)

How can we best challenge the various models for phage recombination? Phage workers may be clever enough to design crosses, with or without density labels or other tricks, that can challenge central aspects of the schemes. More likely, however, the next round of advances will come from in vitro studies on recombination. The smashing successes with the Int system are certain to inspire studies of Rec, Red, and other generalized systems acting in vitro. If a good mimic of the in vivo systems can be achieved, then selective removals and additions of enzymes and intermediates may give us a fairly secure view of the steps actually involved. Such progress is likely even before this book is in its binding.

PROBLEMS

10-1 Propose and defend recombination models for T4 and λ that differ from the ones presented in this chapter. Alternatively, find and evaluate proposals in the literature.

10-2 Try to classify the molecular models in this chapter according to the schemes under which models are classified in the appendix.

10-3 Supercoiling plays an essential role in Int-mediated recombination in λ, and perhaps in Rec recombination as well. Though we lack evidence, we might speculate that all generalized recombination in λ is dependent on supercoiling. For the case of Int (in vitro), the act of exchange does not result in loss of supercoiling, and such could be the case for generalized exchange as well. However, the models we entertained in this chapter appear to call for relaxation of both donor and recipient as a result of the exchange. In λ, unlike the case in *E. coli* and possibly T4, we may suppose that the circular chromosome is either all supercoiled or all relaxed; there is no evidence that portions of it can relax or coil separately from other portions. If such be the case, then an exchange anywhere on λ would interfere totally with exchanges elsewhere "in the same mating." Such a conclusion is at odds with the mating theory analysis of λ we conducted in Chapter 2. That analysis attributed negative interference (not LNI) to multiple exchanges in matings almost randomly distributed with respect to genotype. (A minor deviation from randomness was acknowledged, with the finite-input factor $g = 0.8$.)

If matings are characterized by complete positive interference, from where will the observed negative interference come? Perhaps it comes from other kinds of heterogeneity, such as bad intracellular mixing of clones. This and other possible heterogeneities can be accommodated in the mating theory by letting g describe all heterogeneities, including finite input, and selecting an appropriate value for it.

Write a mapping function for a phage (λ?) in which there is but one exchange per mating ($S = 0$) and in which r reaches a plateau of 0.15 with increasing map distance. Compare your function with that graphed in Figure 2-6, for which it was assumed that $S = 1$.

NOTES

1. F. W. Stahl (1969) *G* **61** (Suppl):1; R. Holliday and H. L. K. Whitehouse (1970) *MGG* **107**:85.

2. M. Gellert, K. Mizuuchi, M. H. O'Dea, and H. A. Nash (1976) *PNAS* **73**:3872.

3. M. S. Meselson and C. M. Radding (1975) *PNAS* **72**:358.

4. M. Mizuuchi and H. A. Nash (1976) *PNAS* **73**:3524.

5. J. Weil and E. R. Signer (1968) *JMB* **34**:273.

6. F. W. Stahl and M. M. Stahl (1977) *G* **86**:715.

7. F. W. Stahl, K. D. McMilin, M. M. Stahl, J. M. Crasemann, and S. Lam (1974) *G* **77**:395.

8. P. V. Sarthy and M. Meselson (1976) *PNAS* **73**:4613.

9. M. S. Meselson (1968) in *Replication and Recombination of Genetic Material*, W. J. Peacock and R. D. Brock, eds., Australian Academy of Science, Canberra, p. 152.

10. R. K. Herman (1965) *JB* **90**:1664.

11. B. J. Thompson, C. Escarmus, B. Parker, W. C. Slater, J. Doniger, I. Tessman, and W. C. Warner (1975) *JMB* **91**:409.

12. H. Potter and D. Dressler (1976) *PNAS* **73**:3000.

13. N. Sigal and B. Alberts (1972) *JMB* **71**:789.

14. M. Valenzuela and R. Inman (1975) *PNAS* **72**:3024.

15. B. J. Thompson, C. Escarmis, B. Parker, W. C. Slater, J. Doniger, I. Tessman, and R. C. Warner (1975) *JMB* **91**:409.

16. B. J. Thompson, C. H. Sussman, and R. C. Warner (1978) *V* **87**:212.

17. B. J. Thompson, M. N. Camien, and R. C. Warner (1976) *PNAS* **73**:2299; H. Potter and D. Dressler (1976) *PNAS* **73**:3000.

18. M. Meselson (1972) *JMB* **71**:795.

19. E. A. Birge and K. B. Low (1974) *JMB* **83**:447.

20. F. W. Stahl and M. M. Stahl (1974) *MGG* **131**:27.

21. W. Gilbert and D. Dressler (1968) *CSHSQB* **33**:473.

22. J. S. Haemer, personal communication.

23. F. W. Stahl, K. D. McMilin, M. M. Stahl, J. M. Crasemann, and S. Lam (1974) *G* **77**:395.

24. F. W. Stahl, M. M. Stahl, and R. E. Malone (1978) *MGG* **159**:207.

25. F. W. Stahl and M. M. Stahl (1974) in *Mechanisms in Recombination* R. F. Grell, ed., New York: Plenum, p. 407.

26. F. W. Stahl, K. D. McMilin, M. M. Stahl, J. M. Crasemann, and S. Lam (1974) *G* **77**:395.

27. D. G. Ross and D. Freifelder (1976) *V* **74**:414.

28. P. V. Sarthy and M. Meselson (1976) *PNAS* **73**:4613.

29. Ibid.

30. K. Mizuuchi and H. A. Nash (1976) *PNAS* **73**:3524.

31. A. Landy and W. Ross (1977) *S* **197**:1147.

32. M. Shulman and M. Gottesman (1973) *JMB* **81**:461.

33. L. W. Enquist, personal communication.

34. T. R. Broker (1973) *JMB* **81**:1.

35. D. Pettijohn and R. Hecht (1973) *CSHSQB* **38**:31; K. Drlica and A. Worcel (1975) *JMB* **98**:393.

36. A. H. Doermann and D. H. Parma (1967) *JCP* **70** (Suppl. 1):147.

37. G. Mosig, R. Ehring, W. Schliewen, and S. Bock (1971) *MGG* **113**:51.

11

Models for
Fungal Recombination

Conversion and exchange in fungi can be related by models similar to those called upon to explain recombination in phage. All of these models invoke annealing of complementary chains to ensure precision of the exchange. They vary, however, with respect to the reciprocality of events and the relative roles of mismatch correction and nonmismatch-stimulated DNA synthesis in conversion.

CLASSIFICATION OF MODELS

So many molecular models for eukaryotic recombination have been put forth that we cannot treat them all. To make sense out of the ones we do examine, we need a classification scheme. Of the many schemes possible, I suggest we use one implied in Chapter 8. That discussion indicates that models may be classified according to the nature of the finished products as they would appear in the absence of mismatch correction or other marker effects. Such a classification scheme divides models into three classes (Figure 11-1). Although some models resist classification by this scheme, I think we will find it useful nevertheless.

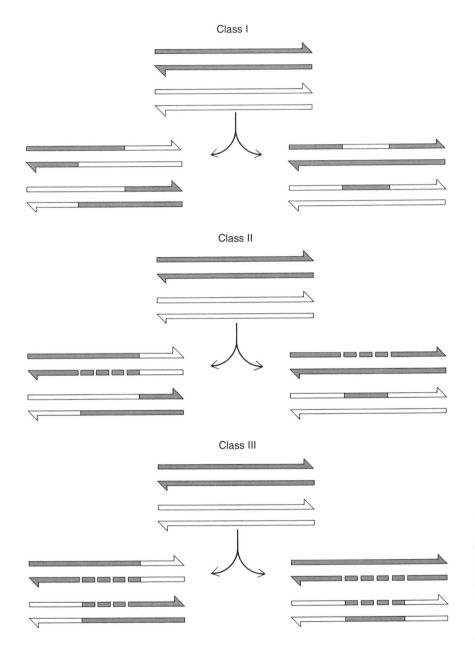

FIGURE 11-1
The three classes of models for eukaryotic recombination based on the nature of the finished products. Class I: No new DNA; reciprocal splatches. Class II: New DNA on one chromatid; splatch on the other. Class III: New DNA on both chromatids; no patches.

CLASS I MODELS

Holliday's model is the prototype of class I models (Figure 11-2). During or after splatch formation, mismatch correction occurs with direction and efficiency that varies with the nature of the mismatch and the organism.

We shall now take the steps in the model one at a time and speculate on their chemistry. In Holliday's formulation the initial cutting occurs at special sequences he calls recombinators. Recombinators are fixed starting points for splatch formation. He likes the idea of a single enzyme molecule with two active sites operating in unison to assure synchronous cutting (the Int system of λ may be an analog of what Holliday had in mind). If *cog* or Chi are recombinators, then Holliday's scheme needs revision at this first

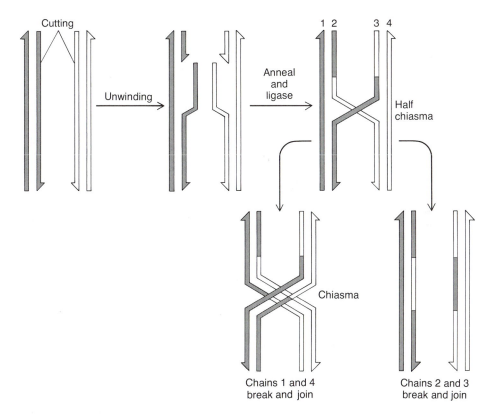

FIGURE 11-2
Holliday's model. (R. Holliday [1964] *GR* **5**:282.)

step. The minimal revision I can think of is that *cog* or Chi are binding sites, but not sites of action for Holliday's cutting enzyme. Once bound to one or the other chromatid, the enzyme slides. At some variable distance from the binding site the enzyme makes the twin cuts of the Holliday model. Whether there is any sequence requirement for this cutting is moot, but it seems simpler to suppose that there is not. Looked at this way, the starting point for splatch formation is variable but is focused about the binding site.

Step 2 of Holliday's model requires melting of the DNA. We might suppose this to be catalyzed by ATP-splitting, DNA-unwinding enzymes like the *recBC*$^+$ product. Single-chain-binding proteins might protect the exposed single chains and facilitate the annealing step. There are two ways of thinking about chain swapping. (1) In its original formulation the model supposed that the length of DNA unwound was equal to the ultimate length of the splatch. (2) A more recently recognized alternative is to suppose that the initially unwound and swapped chains are short; they achieve their final lengths by branch migration, i.e., by "rotary diffusion" of the chromatids. Both schemes have their critics. Some find the idea of long single-chain stretches of flopping DNA distasteful. Others wonder whether paired chromatids have the necessary freedom of motion to enjoy rotary diffusion. The two alternatives do appear to make one differing prediction.

A rather gross genetic difference (e.g., a deletion) might scarcely interfere with splatch formation if the two chains indeed unwind and then anneal over the entire splatch length. This reaction might have essentially the same properties as in vitro annealing of DNA that proceeds effectively between complementary chains from two duplexes that differ by a deletion or a substitution. The rotary diffusion proposal, however, requires that each step in splatch formation make no change in the stability of the structure. Movement of a splatch into a nonhomologous region would create unpaired bases, lowering the stability of the structure. With this proposal, we cannot expect splatches to cover deletions. For T4 we argued that deletion-heteroduplex splatches are formed, and for λ we argued that they are not (Chapter 6). Evidence on this point in fungi is not straightforward. Conversion asci for deletion mutants have been reported for yeast. Should any of these prove to show postmeiotic segregation, we would have proof of deletion-heteroduplex splatches. *If* we accept the correctness of class I models to the exclusion of others, conversion for deletions would itself constitute proof of deletion-heteroduplex splatches, and we would lean toward Holliday's original proposal for the mode of splatch formation.

Resolution of the half-chiasma to give splices, on the one hand, or patches, on the other, can likewise be thought of in several ways. However, any hypothesis must have the property that splices and patches form with comparable probability, although most data say that an edge should be given to patches.

In Holliday's original model the cutting of cross-connecting chains (2 and 3 in Figure 11-2) and noncrossed chains (1 and 4) are mutually exclusive reactions of equal, or comparable, probability. One can imagine one cutting enzyme with two activities or two enzymes acting with fortuitously similar probabilities. A minor modification of this scheme supposes that splatches extend by rotary diffusion until a nick is chanced upon. If the nick is in a cross-connecting chain, diffusion stops and the other cross-connecting chain becomes endonuclease-sensitive as the two duplexes diffuse a bit apart. If the nick is in a noncrossing chain, then the second noncrossing chain becomes endonuclease-sensitive because of loss of rigidity of the intermediate.

A second way of thinking about resolution is to suppose that the neatly arrayed parallel chromatids of our diagrams are mythical and that the intermediate looks instead like the diagram below (I do not mean to imply that chromosomes do not pair in parallel alignment in meiosis, only that the local DNA configurations need not reflect that gross parallelism):

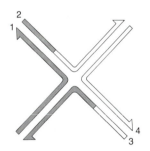

Recall that the cross regions of the figure eights of presumptive prokaryotic recombinational intermediates do look like this in the electron microscope (Chapter 10). Now we may imagine one two-headed enzyme whose sole catalytic ability is to cut such a cross either horizontally or vertically:

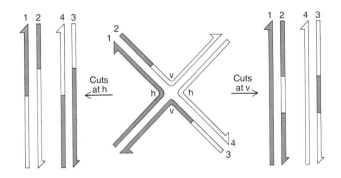

A third view supposes that the above cross structure is but a transient intermediate between two isomeric forms of parallel duplexes in which the crossing and noncrossing chains have swapped places:[1]

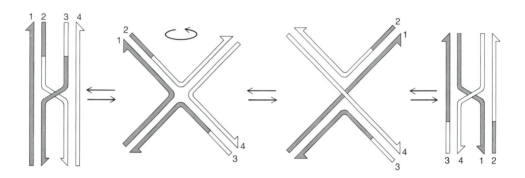

The two isomers with parallel duplexes are subject to the activity of a single kind of endonuclease that cuts the crossed chains.

This scheme has an intriguing feature. The half-chiasma intermediate does a lot of flopping to get from its initial state to the structurally identical isomeric state. Perhaps the flopping requires rather long-range adjustments of the paired chromatids. These long-range consequences might cause interference. We may suppose, for instance, that all chain swapping occurs before any isomerization and is free of any interference. Two nearby half chiasmata might then be sterically forbidden from isomerizing. The results would be interference between splices but no interference between splices and patches or between patches and patches. In fact, one would see no interference between splatches, or, operationally, between conversions. (We considered one version of this model in Problem 8-4.)

The last step in the model is the correction of mismatches.

The model *is* tidy. However, evidence arguing for nonreciprocal splatch formation led to derivative models. Later we shall return to Holliday's model to see how it might be retained in the face of the apparent non-reciprocality of splatches.

CLASS II MODELS

The essence of this class of models is the nonreciprocal move of a chain from one chromatid to its homolog. A corollary of this nonreciprocality is that space must be made to receive the moving chain, and the loss of that chain from its original duplex must be compensated for. These requirements

Models for Fungal Recombination

mean that DNA must be discarded from the recipient duplex and synthesized on the donor duplex. The various models in this class differ according to the sequence of these events and the nature of the causality among them.

The donor is cut on one chain and opens up. The opening could be by a $recBC^+$-like unwinding enzyme aided by single-chain-binding protein. The abandoned chain is then provided with a complement by chain extension:

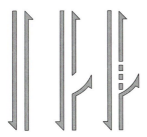

The same end result can be achieved by chain extension itself displacing the cut chain. The recipient chromatid, perhaps because it is supercoiled, accepts this chain:

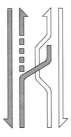

The wronged (displaced) chain is then digested away, giving:

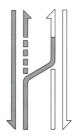

The splatch can be made longer by continuation of chain extension coupled with digestion on the recipient chromatid:

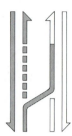

The transfer process could stop in any of several ways. The simplest I can think of is the encountering of a nick on the crossing chain, giving us a nonreciprocal patch:

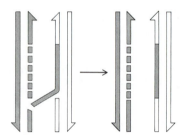

A cute way to bring things to a stop is to allow the chromatids below the crossing chain to swap places:[2]

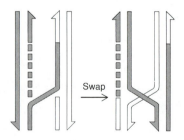

Three-dimensional mock-up models suggest that this may be as easily done as said. Ligase is called upon to seal things up here, as elsewhere. This half-chiasma intermediate can be cut by the hypothetical endonuclease for crossed chains, giving us a nonreciprocal splice with reciprocal exchange of flanking material:

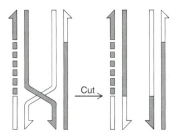

It appears in this diagram as though I made a crossover without a chiasma. In fact, the swapping of duplexes is imagined to be a strictly local event so that the chromatids really look like this after the exchange:

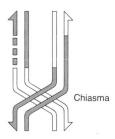

Chiasma

The model as elaborated to this point allows for comparable frequencies of patches and splices but does not demand this.

Now note that following duplex swapping and ligation the resulting half-chiasma intermediate can be supposed to have all (or any part of) the properties we described for class I models. In particular, the half-chiasma may move in either direction by rotary diffusion:

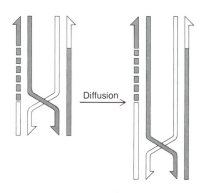

Adoption of that supposition puts one leg of this model into class I—it allows for partially reciprocal splatches in a model that started out non-reciprocal![3] The model is sometimes called the Aviemore model after the village in Scotland where it was conceived.

If the symmetric, diffusion phase of the Aviemore model is long-lived compared to the asymmetric phase, predictions of the model converge with those of class I.

EVENTS START ON THE RECIPIENT

The recipient is nicked, perhaps at a recombinator (like *cog*, Chi, or *M26*). The nick will have a longer life if we enlarge it to a gap. So be it. The gap qualifies the chromatid as a recipient. (If we like, we can suppose that the gaps are a residuum of DNA segments replicated on only one chain at the premeiotic DNA replication.[4]) In order to get the homolog to donate, we let one end of the gap be opened up, e.g., with a *recBC*+-like unwinding enzyme. The single-chain whisker can be taken up by a supercoiled homolog: The wronged (displaced) chain of the supercoil can now be nicked and transferred into the gap:

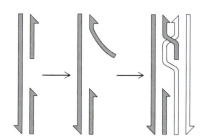

Chain extension on the donor and trimming as necessary brings us to a familiar intermediate:

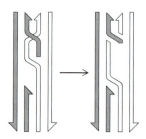

Models for Fungal Recombination

The model then continues like the one in which events start on the donor.

CLASS III MODELS

Models of this class[5] require local replication of at least one of the two participating chromatids.

CHAIN EXCHANGE INITIATES REPLICATION

Our first class III model begins much like others we have considered. A melted-out chain, nicked at a fixed place, is taken up by a duplex, perhaps aided by supercoiling:

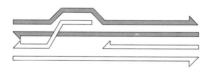

The 3′ end then primes chain elongation:

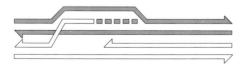

and a replication fork develops:

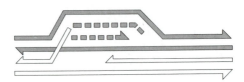

Degradation of the 5′-ended chain gives:

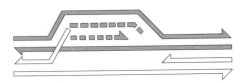

Any of several sequences of events can result in an exchange at the replicating fork. One way to think about this is to suppose that the stretch of single-chain DNA at the fork is nuclease-sensitive. Cutting there and then joining black to white gives:

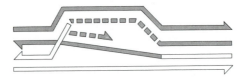

Now let us banish the segment of single chain on the white duplex (by again invoking single-chain specific nuclease) and join chain ends of proper polarity:

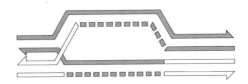

Completion of the synthesis and discard steps has led us to a Holliday structure. Resolution can proceed as in Holliday's model to give either parental or recombinant arrangements of flanking material. Prompt resolution gives us class III recombinants with a bit of splatch where the event originated. Branch migration prior to resolution would give us a model that is hybrid between class III and class I. The unique feature of class III models is that 6:2 tetrads can arise without mismatch correction.

Sex Circle Models

Models like those described above have their attractions and adherents. However, they do not handle in a totally satisfactory way one very good body of data, namely that collected by full analysis of random tetrads in yeast.[6] The features of the yeast data that put special demands on models are the following.

1. Any pair of markers can recombine either reciprocally or by conversion.
2. The closer the markers the larger the fraction of nonreciprocal recombination events.
3. Conversion shows "parity" for (almost) all markers, including large deletions; i.e., conversion to mutant and to wild type is equally frequent for each marker.
4. Conversion shows "polarity." There is a position-dependent hierarchy in conversion rates, which suggests the use of special places, such as borders of fixed pairing regions or starting points, for splatch formation.
5. Among convertants, flanking markers are recombined about half the time.
6. Among reciprocal recombinants, flanking markers are recombined 100 percent of the time.
7. Conversion that does not recombine flanking markers sets up no interference in an adjacent interval.
8. Conversion that recombines flanking markers sets up interference in adjacent intervals. Reciprocal recombination of very close markers also sets up interference.
9. 5:3 and 6:2 tetrads behave identically with respect to the constraints listed above.
10. Postmeiotic segregation can occur for almost any marker; it has not been reported for deletions. It varies for different markers but is usually a small fraction of all conversions.

11. Aberrant 4:4 tetrads are very rare and may well represent the consequences of two separate events. Congruently, there is no other evidence that events manifested as conversions are ever reciprocal or involve the two participating chromatids in any symmetric intermediate.

This is a remarkable set of constraints. As we noted earlier, these constraints are not applicable to recombination in all fungi, but they do deserve special attention in our model-building efforts. Only one model, as far as I know, has comprehended conversion in yeast.[7] That model is not very molecular; it is more a geometric restatement of the facts of recombination in yeast.

The sex circle model is a fixed pairing segment model. In the modified form described here (call it the "unisex circle model") the model supposes a localized replication on one of the two interacting chromatids:

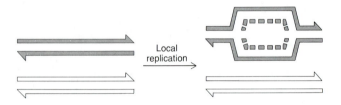

The replication starts from a fixed point and proceeds in one or both directions. The termination points may be fixed or variable. We then suppose the circle to be recombinogenic (the model does not specify how, but it is not too hard to cook up molecular possibilities). Each circle then enjoys two splicings with the other chromatid. The two exchanges may involve the same arc of the circle or the two different arcs with equal probability:

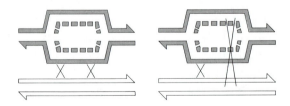

The exchanges are nonreciprocal; one of the two products of each exchange is destroyed, giving:

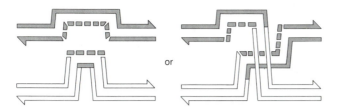

When the kinks are pulled out of the diagram we have:

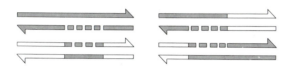

Now reexamine the list of characteristics of yeast recombination and convince yourself that the model accounts for each item on the list or is easily compatible with it. I find that the model can even account for the phenotype of the *rec4* mutation described in Chapter 9. For it to do so we need only suppose that the *rec4* mutation permits the *arg4* sex circle to expand further than is normal. With this assumption alone the model accounts for the reduction in reciprocal exchange between very close markers and the increase in (mean) length of conversion segment while introducing little, if any, change in conversion rate.

In checking the model against the list of constraints imposed by the data, you may have had trouble with item 10: marker-to-marker variability in the fraction of conversions that show postmeiotic segregation. I can accommodate it by supposing that the location of exchanges within sex circles may be nonrandom for reasons having nothing to do with the geometric features of the model. A second possibility is that either sex circles or splices have different mean lengths in different parts of the chromosome. Of course, if need be, the model could allow a bit of mismatch correction or other marker effects.

Interference for intervals within the circle follows from the feature that each circle has exactly two exchanges, and both are used up in making a crossover. For interference beyond the circles some additional source will have to be sought.

There is one awkward feature of the modified sex circle model described above and its original version. Care must be taken that the two nonreciprocal break-join events do not carelessly discard DNA, which would leave the arms flanking the circle unable to rejoin. This awkwardness is avoided in a version of the model constructed for the purpose. Known as the molecular sex circle model, this version has the added virtue of being a good deal more "molecular" than the unisex circle model or its predecessor.

We suppose that an exchange occurs at, and is stimulated by, the replicating fork of a sex circle initiated at a fixed place. The sequence of events described below is one way to imagine this:

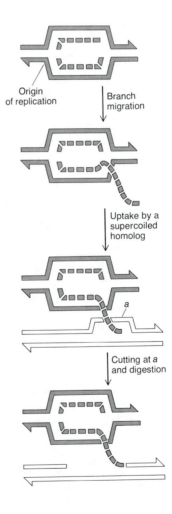

The free end at the left of the gap can now "attack" the circle, either on the lower arc (leading to conversion without exchange of flanking material)

or on the upper arc (leading to crossing-over). I illustrate the former case because it is easier to do so:

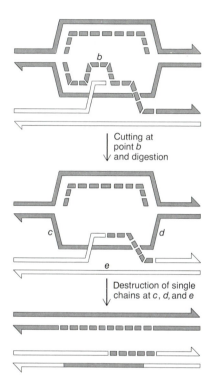

Cutting at
point *b*
and digestion

Destruction of single
chains at *c, d,* and *e*

You may think up a neater sequence of events. The essential feature is that the first event is intrinsically asymmetric and that the asymmetry in the second event results from the asymmetry of the first.

MODELS VERSUS DATA

There is nothing terribly attractive about the sex circle models even in their molecular versions, and it may be that the data that prompted them can be easily enough accommodated in some more appealing scheme. The Aviemore model, with its porridge of asymmetric and symmetric splatches plus correction-as-you-like-it, is a versatile candidate. Data like those in Figure 7-1 challenge the models. Note, for example, the 58 tetrads that show 6:2 segregation at *arg4–17* but give normal 4:4 at all the other sites.

Of the 58, 34 are crossover tetrads for the flanking markers, *arg4–16* and *thr*. How are these tetrads accounted for on the sex circle and Aviemore schemes? Let us try the sex circle model first, in its modified ("unisex") version.

The polarity in the data of Figure 7-1 (by polarity I now mean the declining rate of conversion of the *arg4* sites, reading from right to left) indicates a special place either to the right or left of *arg4*. In the unisex sex circle model with fixed ends, the high rate of conversion for the right-hand markers means that there is a fixed end to the sex circle not too far left of *arg4–19*, while the other end of the circle is far to the right of *arg4–17*. The 58 conversions confined to *arg4–17* imply one sex-circle exchange between *arg4–16* and *arg4–17*, with the other sex-circle exchange in the relatively large expanse of circle to the right of *arg4–17*. About half the time ($\frac{34}{58}$) the two exchanges involve opposite arcs of the circle and thereby reciprocally recombine the flanking markers. (Note the seven cases in which one of the exchanges landed on *arg4–16*, giving a 5:3 segregation there.) I find the other major classes of tetrads to be as easily accommodated by the model. How does the Aviemore model fare? We shall examine two versions of it.

CONVERSION HIGH NEAR THE INITIATOR

In this version of Aviemore the high rate of conversion for sites on the right side of *arg4* implies an initiator to the right of *arg4*. The asymmetric Aviemore event is initiated there and invades *arg4* leftward.

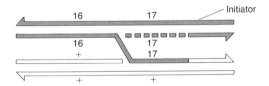

(I have arbitrarily drawn the case in which initiation occurred on the *4–17* mutant chromatid, and for which conversion will result in tetrads with an excess of *4–17* mutant information). Some of the events diagrammed above go undetected because the structure is resolved to an asymmetric patch, and the patch gets corrected to wild type. However, let us focus on those cases in which isomerization occurs in the *16–17* interval. Since there is little or no symmetric splatching in yeast (Chapter 8), resolution must follow upon isomerization, with little or no branch migration by rotary diffusion. Most of the 58 conversions involving only *arg4–17* must have terminated in the *16–17* interval. Thirty-four of them were resolved after isomerization (so as to reciprocally recombine flanking markers) and were corrected on the recipient chain. If parity in yeast has as its explanation that correction is always on a recipient chain, then this version of the Aviemore model appears competent. If we do not adopt the rule that correction is always on the recipient chain, however, this version is in trouble. For the 34 splices with correction of *17* to mutant, there should have been, because of parity, another 34 corrected to wild type. These tetrads would appear simply as crossovers between sites *16* and *17*, with no sites converted. However, only one such tetrad was seen, far less than the 34 expected.

If we do apply the rule that correction is always on the recipient chain, we shall have trouble accounting for other features of the data in Figure 7-1. Note the 13 single-site conversions at site *19* and the 10 at site *16*. Each of these conversions was accompanied by normal segregation at sites to the right. (At site *16*, segregation was normal at sites on both sides.) If an asymmetric splatch begins at the right of *arg4–17* and invades *arg4* leftward (to account for the polarity), then conversion at site *16*, for instance, must involve splatch formation across *17*. In that case, why is there no abnormal segregation at *17*? Because, one must suppose, *17* was corrected back to 4:4. But that supposition violates the previous assumption that correction is always on the recipient chain! Can this version of the Aviemore model be saved? It can if one is willing to loosen the rules so that initiation can occur anywhere throughout the gene, e.g., between sites *16* and *17*. The overall polarity in the data would then be due to an especially strong initiator right of *17*.

Conversion High Near a Terminator

Perhaps the feature that accounts for a high conversion rate at the right end of the *arg4* gene is a nearby "terminator" of an Aviemore reaction.[8] We may imagine an initiator left of *arg4*. The initiator has similarities to *cog* and Chi; i.e., it acts at a distance. We suppose an enzyme binds there and then either slides, gaps, or translates a nick rightward. At any moment the moving enzyme may begin a rightward-moving asymmetric splatch. The splatch continues to the terminator (which may lie between two genes and not within a gene) where it can resolve as a splice or a patch with comparable probabilities. The few (9 + 1) fully reciprocal events seen within *arg4* may reveal a small probability of termination before the terminator is reached. We diagram the 58 tetrads that are 6:2 for *arg4–17* this way:

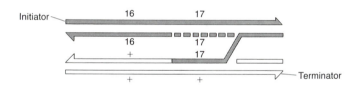

If correction of the heteroduplex is toward $+$, then there are no observable consequences for *arg4*. If correction is to *arg4–17*, then the events diagrammed are the 58 that give $6:2$ segregation at that site. Of these, 34 are resolved in the isomerized state at the terminator. This version fits this datum, but does it fit the rest of the data in Figure 7-1? It appears to me that it works, but only if correction is highly probable (to account for the low frequency of conversions showing postmeiotic segregation) and if excision tracts are long (to account for the absolute lack of cases in which *19* and *17* are corrected to give a conversion while the middle site, *16*, is corrected to give a normal $4:4$ segregation). That all seems reasonable enough; the only part I find a little hard to take is the idea that deletions should show parity due to the equal probability of correction of heteroduplexes in the two directions.

The two versions of the Aviemore model presented above differ with regard to the location of the conversion-associated chromatid exchange. In the first version it often occurs within the gene we are watching, while in the second version it occurs outside of it, at the high-conversion end. For a single-site conversion, the first version demands that chromatid exchange be on the low-conversion side of the site, while the second version puts it on the high-conversion side. An ingenious use of $5:3$ asci has provided information on this matter.[9]

Asci that gave $5:3$ segregation at site *16* in *arg4* of yeast were scored for crossing-over of flanking markers at *pet3* on the left and *thr* on the right. Different kinds of $5:3$ asci can be distinguished in the cross

pet+	16	thr+
	$\times$	
pet	16+	thr

We illustrate by showing the two $5^m:3^+$ classes:

<table>
<tr><td>pet+ 16 thr+
pet+ 16 thr+</td><td rowspan="4">and</td><td>pet+ 16 thr+
pet+ 16 thr+</td></tr>
<tr><td>pet 16 thr+
pet 16 thr+</td><td>pet+ 16 thr
pet+ 16 thr</td></tr>
<tr><td>pet+ 16 thr
pet+ 16+ thr</td><td>pet 16 thr+
pet 16+ thr+</td></tr>
<tr><td>pet 16+ thr
pet 16+ thr</td><td>pet 16+ thr
pet 16+ thr</td></tr>
</table>

Convince yourself that the left-hand ascus has experienced an exchange to the left of the converted site while the right-hand one has its exchange to the right. Both versions of the Aviemore model are embarrassed by the observation of associated exchange on *both* sides. There were, in fact, about 20 asci exchanged to the left and 44 to the right of *arg4–16*. The various sex circle models have no difficulty with this datum. Consider the unisex version. A 5:3 segregation at site *16* indicates that one of the two exchanges postulated to occur within the circle occurred on site *16* via a splice. The other occurred left or right of that, depending on the relative space available.

Recent data on conversion in *Ascoblus*,[10] however, fit the Aviemore model. In one-factor crosses, asci were analyzed for each of a number of markers arrayed along a gene. The frequency of aberrant asci decreased from left to right. The mutants used included the two types described in Chapter 7. Among mutants that gave appreciable frequencies of post-meiotic segregation, the proportion of 4:4 aberrants to total aberrant asci increased from left to right on the gene. This is evidently in accord with the first version of the Aviemore model, in which splatches initiated asymmetrically at the high-conversion end of a set of markers become progressively symmetric further away. Those mutants that gave low frequencies of post-meiotic segregation and large disparities in the direction of conversion also manifested a gradient in accord with the Aviemore model. For these the gradient was in the magnitude of the disparity, with the disparity becoming greater toward the low-conversion end of the gene. The change in disparity signals a change in the relative frequencies of two conversion mechanisms. At the left end of the gene, extra copies of markers are the result of both the DNA synthesis that accompanies chain transfer from one chromatid to the other and mismatch correction. At the right end, extra copies are primarily a result of mismatch correction. Since conversion disparity presumably reflects bias in the direction of mismatch correction, disparity is strong when most conversion is a result of correction and weaker when part of the conversion is by a mechanism expected to manifest parity.

SPLATCHES BY CORRECTION

The concept of splatches has been central to our discussions. Indeed, the value of the splatch concept for thinking about the precise alignment of parental contributions to recombinants is overwhelming. Splatches are detected in crosses by detecting postmeiotic segregation and, if you are misguidedly sold on the ubiquity of mismatch correction, by detection of 6:2's as well. The splatches we drew in the molecular models of this chapter, were, in each case, composed of material as well as information derived from the two different parents. However, some of the splatches observed as postmeiotic segregations may have a different origin and significance than we have assumed.

Let us illustrate with the Holliday model.[11] Suppose, for example, that the half-chiasma intermediate sometimes (often?) resolves simply by diffusion back to the recombinators that, by chance, have failed to be ligated. During the lifetime of the intermediate, mismatch correction might have occurred on one, the other, or both heterduplexes:

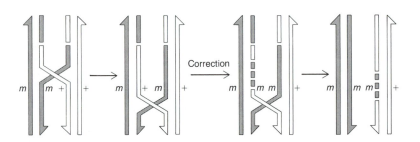

True, the scheme does have incipient material splatches temporarily, but the finished product has a patch that is a *copy* of a stretch of the homologous chromosome. Note that heterozygosity in this patch is not a consequence of the failure of mismatch correction but of the occurrence of it. In fact, the patch itself is a result of correction (with a finite tract) and as such is itself a marker effect!

If correction probabilities are small and these abortive intermediates short-lived, the observed splatches would generally be nonreciprocal even though the intermediate was itself a result of symmetric chain exchange. By pushing these ideas, one might concoct nonreciprocal splatches (by correction) even for the case where the symmetric chain exchange is resolved with recombination of flanking meterial.[12] Try it!

A number of the steps proposed in the models in Chapters 10 and 11 are predicated on the double-helix structure for DNA. In order to bring complementary chains into pairing, we invoked unwinding, degradation of one chain, or chain migration by rotary diffusion. However, the helical structure for DNA has been challenged; a structure in which the two chains are related by short helical segments of opposite sense appears to satisfy the known physical and chemical properties of double-chained DNA.[13] Should this heretical model prevail, our models might change in the direction of greater simplicity.

I was hoping that by this point I would have some good ideas on the possible role of premeiotic DNA replication on recombination. There is no one idea I am fond of, but we might get some clues from the phages. First, the ability of Rec and Int to recombine unreplicated λ advises us that recombination can occur without immediately prior chromosome replication.

However, the ubiquity with which meiotic recombination is preceded by chromosome replication is too provocative to ignore.

If we knew the molecular mechanism by which replication enhances the action of Red on λ, we might have a clue to the role of premeiotic replication in recombination. For T4 we have speculated that replication provides terminal single-chained regions as a result of asymmetry of the replicating fork. Perhaps premeiotic replication supplies recombinogenic single-chained regions in a similar fashion.

The models we examined in this chapter did not include any in which the initial step is a chop, i.e., an enzymatic cut of both chains, severing a duplex. Such models deserve consideration, however, because of the evidence that X-ray-induced chops are recombinogenic as well as the evidence that T4 ends are recombinogenic. I excluded such models because I have not seen any I liked. The shortage of attractive chop models may reflect a general unwillingness to consider them seriously. They are a priori unattractive—if a duplex is chopped, what is to keep the two halves from floating off? But here is a possible role for premeiotic replication! A chromatid could act as a scaffold for its sister, holding pieces together after the chop. This role could best be played if premeiotic replication were partial, as in the model Mark IIB (p. 42). The scaffold model might also be parlayed into an explanation for the absence of (or peculiar properties of) sister exchanges.

Perhaps DNA-modifying enzymes (methylases, etc.) are absent in premeiotic cells. If so, premeiotic replication would provide DNA chains that remain unmodified. These unmodified chains might be uniquely sensitive to nucleases that initiate recombination. Differential modification of chains might be involved in correction, too. A heterozygous splatch that contained one modified and one unmodified chain might be corrected preferentially on one chain or the other. If DNA modification played a role both in exchange and correction, then schemes in which correction occurred preferentially on recipient chains would be easy to devise. We toyed with such possibilities earlier.

One can suppose that premeiotic DNA replication plays no mechanistic role in recombination and try to explain its ubiquity instead on teleological grounds. Could it be that recombination in the four-strand stage is a kind of conservatism? Recombination (presumably) occurs to generate variants. But for any single diploid cell, the generation of variants is a risky step; they may be ill-adapted. If crossing-over were at the two-strand stage, each diploid cell would risk all or nothing. For a single-celled eukaryote, like yeast, this might be serious. Exchange at the four-strand stage, however, allows the hedging of evolutionary bets. A lot of thought is needed on the significance of crossing-over at the four-strand stage. The role(s) of DNA synthesis in prophase of meiosis needs examination, too.

Perhaps pachytene synthesis (observed in lilies) is needed to prepare chromosomes for exchange, as Red seems to need replication prior to recombination. The sex circle models build on such a possibility. Or perhaps pachytene synthesis represents the repair of gaps that either initiate or terminate an exchange, as we have speculated for Chi-stimulated Rec recombination. Possibly, pachytene synthesis is only a reflection of mismatch correction in cloned hybrids.

We have struggled at length with the origins of negative interference. What, however, is the cause of positive interference? And why do splices have it but patches (or paired splices?) not? The models in this chapter have made splices and patches alternate outcomes of a single branched reaction sequence. Is that the best way to think about it? Maybe, but we would all feel better if a good molecular explanation for the facts of interference were neatly wedded to some of those models.

PROBLEMS

11-1   This problem expands on Problem 8-3.

Two authors[14] derive a model for splatches of fixed length with splices and patches equally frequent (Equation A-5 and Figure A-6, in the appendix, with $a = 0.5$). Their starting point for the derivation is the Holliday model in pure form. They then wonder about the consequences of modifying the Holliday model by relaxing the feature of "... formation of hybrid DNA [incipient splatches] on both participating chromatids. There are reasons for thinking that this might not always be the case, since repair of mismatched bases in hybrid DNA on both chromatids should yield reciprocal intragenic recombinants, and in many sets of data these are not observed. If hybrid DNA is nonreciprocal with a unilateral donation of a stretch of single strand [i.e., single chain] from one chromatid to the other and with the doublestranded structure of the donor chromatid being restored by repair-copying from its own surviving single strand, this will effectively half the yield of recombinants [at $D/L > 1$]... without affecting the yield of recombinants from reciprocal crossovers resulting from hybrid DNA falling *between* the sites. If hybrid DNA formation is *always* nonreciprocal, and [$a = 0.5$], ... the initial and final slopes should be the same."

We know that a linear dependence of R and D is obtained only when the coefficient of coincidence is zero. We also know that patches are a source of localized negative interference; in their presence it is inconceivable (barring compensating marker effects, such as map expansion) that the coefficient of coincidence will be zero. There must be a fallacy in the argument of the two authors.

(a) Derive a relationship between R and D (or between R/kL and D/L) that correctly describes the nonreciprocal-hybrid-DNA Model.

(b) What is the ratio of the final to the initial slope in your model?

11-2 Ignoring mismatch correction, which of the mapping functions encountered previously or in the appendix apply most closely to the class I, II, and III models of this chapter?

11-3 In the framework of class I, II, and III models, respectively, how might you explain the consequences of the *rec4* mutation in yeast (Chapter 9)?

11-4 At least one model for eukaryotic recombination has been offered in which the initial event is a double-chain break of one chromatid. See if you can fashion a pleasing model on that assumption. Identify those eukaryotic phenomena that your model accounts for and those it does not.

11-5 Write a mapping function for very close marker recombination according to the unisex circle model. Assume both ends of the sex circle to be fixed at points outside the marked interval, and let the marked interval be centrally placed in the circle.

11-6 Consider two markers near each other and both close to the left end of a unisex circle with two fixed ends.

(a) Which marker will manifest the higher overall rate of conversion?

(b) Among tetrads containing recombinants for the two markers, which will manifest the higher rate of conversion?

(c) Among tetrads recombinant by virtue of conversion at the right-hand site, What fraction will be tetratype for markers flanking the sex circle?

(d) Among tetrads recombinant by virtue of conversions at the left-hand site, what fraction will be tetratype for markers closely flanking the sex circle?

(e) Among tetrads that are co-converted for the two marked sites, what fraction are tetratype for markers closely flanking the sex circle?

(f) Among tetrads reciprocally recombinant for the two close markers, what fraction will be tetratype for markers closely flanking the sex circle?

(g) Among tetrads reciprocally recombinant for the two close markers, what fraction will be tetratype for markers within the sex circle and closely flanking the two close markers? (The appendix provides exercises in this sort of problem and the one following.)

11-7 Our several models for fungi recombination each imply particular relationships between conversion and crossing-over. These relationships may provide tests of the suitability of a given model for a given creature. Let us focus on two simple, highly contrasting models—Holliday's model and the unisex circle model, each formulated in a simple version. For Holliday's model, let the initiation of splatches be at fixed spots and let the distances from initiation to resolutions be exponentially distributed sometimes to one side and sometimes to the other. For the unisex circle model, let both ends of each circle be at fixed positions.

(a) In Holliday's model, how does conversion rate vary as a function of position between two fixed starting points. For the unisex circle model, how does conversion rate vary as a function of position in the circle?

(b) Consider two markers very close together compared to the mean splatch length or the length of the sex circle. For Holliday's model, suppose that excision tracts are fixed in length and longer than the distance between the two markers and that correction is highly probable and occurs at equal rates in both directions. The cross is:

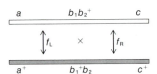

where f_L and f_R mark fixed starting points in Holliday's model or the ends of the circle in the unisex circle model. As usual, the intervals AB and BC are smallish. If sites 1 and 2 are closer to f_L than to f_R, what will be the two most common types of conversion tetrads containing a $b_1{}^+b_2{}^+$ spore in Holliday's model? In the unisex circle model?

11-8 We have used "polarity" in two senses. (1) Within one class of recombinants for two very close markers, the two parental types for flanking markers are unequal. (2) For a set of very close markers, the rate of conversion varies monotonically with position, as in Figure 7-1. Convince yourself that these two phenomena are manifestations of a single process, so that use of a single word is sensible.

11-9 In Chapter 10 we suggested that Chi might resolve figure eights by initiating gaps that run from Chi to the Holliday junction. If such a gap were to pass through a splatch, it would remove any heterozygosity. Since this hypothetical reaction is not stimulated by mismatches, nor is it in any other sense a marker effect, we shall give it a name distinctly different from "mismatch correction." We shall call it "resolution gapping." Since resolution gapping allows for the removal of information and its replacement by allelic information, it is a potential contributor to conversion. Try to build a model for yeast featuring resolution gapping. Start your efforts by imposing the gapping first on Holliday's model and then on various versions of the Aviemore model. I tried this, but found none that I like well enough to tell you about.

11-10 The following passage and table of data are in a 1974 paper (R. Holliday, G **78**:273). Do you agree with the author's contention that "these marker effects are expected if mismatches are removed by an excision repair process"?

The effect of markers is seen by comparing the recombination frequency between two adjacent sites, *arg4-1* and *arg4-2*, in the presence and absence of other *arg4* alleles proximal to this interval. In the 2-point cross, recombination frequency is 1.84 percent, but when mutants *arg4–16* and *17* are also present in the cross, the recombination between *arg* alleles *1* and *2* is reduced to 0.72 percent [see Table 11-1]. This effect could be entirely due to co-conversion of these alleles, yielding no recombinants, triggered by the nearby *16* and/or *17* heterozygous sites. The proportion of co-conversions for *1* and *2* is clearly much greater in the 4-point cross than the 2-point one [see Table 11-1]. Comparable behavior of the alleles *16* and *17* is seen in the presence of either the allele *19*, which is some distance away, or the alleles *2* and *1*, which are much closer. In the latter case, recombination between *16* and *17* is reduced 2.3-fold and the co-conversion frequency increased relative to single-site conversion [see Table 11-1]. These marker effects are expected if mismatches are removed by an excision repair process.

TABLE 11-1
Recombination and conversion at the *arg4* locus in yeast: the effect of nearby sites on recombination and conversion of adjacent mutants *arg4–1* and *4–2* and *arg4–16* and *4–17*.

Cross	Number of tetrads	Conversion of *1* and/or *2*			Crossovers between *1* and *2*	Total recombination between *1* and *2*
		Single-site	Co-conversion	Ratio		
$\dfrac{1\ +}{+\ 2}$	483	25	24	1.04	5	1.81%
$\dfrac{1\ 2\ +\ +}{+\ +\ 16\ 17}$	1,505	39	76	0.51	3	0.75

Cross		Conversion of *16* and/or *17*			Crossovers between *16* and *17*	Total recombination between *16* and *17*
		Single-site	Co-conversion	Ratio		
$\dfrac{19\ +\ 17^*}{+\ 16\ +}$	2,566	69	124	0.56	10	0.87%
$\dfrac{1\ 2\ +\ +^*}{+\ +\ 16\ 17}$	1,505	23	105	0.22	0	0.38

* In these crosses the recombination between adjacent alleles *16* and *19* is 1.61 percent and between *16* and *2* is 0.03 percent.

NOTES

1. S. Emerson (1969) in *Genetic Organization*, Vol. 1, E. W. Caspari and A. W. Ravin, eds., New York: Academic Press, p. 267; N. Sigal and B. Alberts (1972) *JMB* **71**:789; H. M. Sobell (1974) in *Mechanisms in Recombination*, R. F. Grell, ed., New York: Plenum, p. 433.

2. M. S. Meselson and C. M. Radding (1975) *PNAS* **72**:358.

3. Ibid.

4. P. J. Hastings, (1973) *GR* **20**:253.

5. T. Boon and N. D. Zinder (1969) *PNAS* **64**:573; A. Pasewski (1970) *GR* **15**:55; F. W. Stahl (1969) *G* **61** (Suppl):1.

6. R. K. Mortimer and S. Fogel (1974) in *Mechanisms in Recombination*, R. F. Grell, ed., New York: Plenum, p. 263.

7. F. W. Stahl (1969) *The Mechanics of Inheritance*, 2nd ed., Prentice-Hall; F. W. Stahl (1969) *G* **61** (Suppl):1.

8. P. J. Hastings, personal communication.

9. S. Fogel, R. Mortimer, K. Lusnak, and F. Tavares, *CSHSQB*, **43**, in press.

10. N. Paquette, and J.-L. Rossignol, (1978) *MGG* **163**:313.

11. R. Holliday (1974) *G* **78**, 273.

12. Ibid.

13. G. A. Rodley, R. S. Scobie, R. H. T. Bates and R. M. Lewitt (1976) *PNAS* **73**:2959.

14. J. R. S. Fincham and R. Holliday (1970) *MGG* **109**:309.

12

Postscript

Is there one recombination mechanism? Evidently not. Little λ is subject to four generalized systems and one specialized system, and these systems appear to fall into distinguishable classes with regard to reciprocality, site-specificity, and the involvement of chromosome replication and of chain extension. When we look at T4 we see evidence of a recombination scheme that may be different from any in λ. Will the situation in fungi prove as complex? Probably.

The regional specificity of the *rec* mutants of *Neurospora* and the *rec4* mutant of yeast and similar observations on recombinational variants of *Schizophyllum*[1] suggest a number of coexisting recombination systems in each fungus. These systems may be almost identical, differing only in their recognition of different recombinators, i.e., *cog-* or Chi-like signals. On the other hand, they may differ from each other appreciably. Locus-to-locus variation in the results of studies on conversion and exchange require that we acknowledge this.

If a single fungus can host a variety of systems, then it seems inevitable that different fungi will have different, perhaps rather profoundly different, recombination mechanisms.

Is the possibility of diverse mechanisms discouraging? Somewhat. However, the similarities in recombination in creatures as diverse as the phage and fungi are impressive. We can hope that there is but a handful of basic reactions dictated by the physical and chemical limitations of the substrate and that the diversity we see represents but differing mixes of these basic components.

The ubiquity and variety of recombination argues, by the Great Darwinian Tautology, for its adaptive significance. What might that be? One role for recombination seems so obvious that the mention of it strikes me as redundant. Recombination is a source of hereditary variety. That hereditary variety is an essence of life is explicit in the Darwinian view. Is that the only role of recombination? Obviously not. In these pages we have described roles that recombination plays in replication and maturation in phage, and the intimacy with which recombination and reproduction are related in eukaryotes is obvious. That need not detract, however, from the primary role of recombination. At worst, it may merely attest to nature's well-known economies. On the other hand, making reproduction contingent upon exchange, or upon the opportunity for exchange, may further attest to the importance of this process to life itself. In this light, the understanding of recombination is not only an intellectual challenge, it is basic to an understanding of life.

NOTE

1.   G. Simchen and J. Stamberg (1969) *N***222**: 329.

Appendix: Mapping Functions for Very Close Markers

In this appendix we commit to algebraic form the concepts introduced in Chapter 6. Two purposes will be served. First, to some readers the uncompromising nature of an algebraic formalism conveys an idea with more force or precision than does a verbal description. Second, numerical evaluation of our algebraic expressions will help us get a feeling for the quantitative consequences of the various assumptions in the different models.

In the first two sections we examine models in which the probability of recombination is uniform along the chromosome, and marker effects are denied. In the third section we examine mismatch correction operating on uniformly located splices. For each of the models described, we write a mapping function, which shows the dependence of recombination frequency upon distance, and we calculate the coefficients of coincidence as a function of recombination frequency. In the fourth and fifth sections we examine fixed pairing segment and fixed starting point models. In these sections we calculate the frequencies of genotypes for flanking markers among one class of recombinants for two markers covered by the fixed element. In the final section we examine a model in which mismatch correction is imposed upon a fixed starting point model. There we calculate a mapping function, a dependence of S upon R, and the frequencies of genotypes for flanking markers.

EFFECTIVE PAIRING REGIONS

These models account for LNI by supposing that chromosomes pair "effectively" only over short lengths.[1] Effective pairing regions (EPRs) can occur anywhere along the chromosome, but the probability that any particular point will be effectively paired is small. More than one exchange can occur within an EPR. In the simple EPR models, exchanges are Poisson-distributed among EPRs; recombination results when an odd number of exchanges falls between two marked sites. We shall write expressions for two versions: (1) EPRs are of constant length, and (2) EPRs have an exponential length distribution. There are some good reasons for selecting these two versions. First, they represent rather extreme, contrasting possibilities and as such are apt to bracket any interesting situation. Second, they are relatively easy to derive.

EPRs of Fixed Length[2]

We must first define several symbols: D is distance between two marked sites, and if we assume that the chromosome behaves uniformly, D can be measured in (arbitrary) physical units; L is the length of EPRs; and m is the mean number of exchanges per EPR

In deriving our mapping function, we divide our problem into two parts corresponding to $D < L$ and $D > L$, respectively. For $D < L$ we then subdivide into three cases:

CASE I OF $D < L$. The right end of an EPR falls between the marked sites. With probability $k(dx)$ it lies within the interval dx located a distance x from the left site, like this:

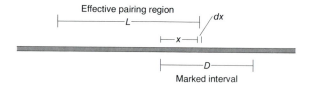

CASE II OF $D < L$. The marked interval falls entirely within an EPR:

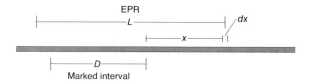

CASE III OF $D > L$.   The left end of an EPR falls between the marked sites. This case is the mirror-image of case I, and we need not solve it separately.

Making use of Haldane's Equation (Equation 1-2) for the probability of an odd number of exchanges, we can write for case I:

$$R_I = \frac{k}{2} \int_{x=0}^{D} (1 - e^{-2mx/L}) \, dx = \frac{kL}{2} \left[\frac{D}{L} - \frac{1}{2m}(1 - e^{-2mD/L}) \right]$$

and for case II:

$$R_{II} = \frac{k}{2} \int_{x=0}^{L-D} (1 - e^{-2mD/L}) \, dx = \frac{kL}{2} \left(1 - \frac{D}{L} \right)(1 - e^{-2mD/L})$$

and case III:

$$R_{III} = R_I$$

Adding cases I, II, and III gives

$$R_{D<L} = kL \left[\frac{D}{L} - \frac{1}{2m}(1 - e^{-2mD/L}) + \frac{1}{2}\left(1 - \frac{D}{L} \right)(1 - e^{-2mD/L}) \right] \quad \text{(A-1a)}$$

The second part of our problem considers $D > L$, and again we have three cases, analogous to the three cases above.

Case I of $D > L$:

$$R_I = \frac{k}{2} \int_{x=0}^{L} (1 - e^{-2mx/L}) \, dx = \frac{kL}{2} \left[1 - \frac{1}{2m}(1 - e^{-2m}) \right]$$

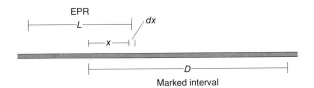

Case II of $D > L$:

$$R_{II} = \frac{k}{2} \int_{x=L}^{D} (1 - e^{-2m}) \, dx = \frac{kL}{2} \left(\frac{D}{L} - 1 \right)(1 - e^{-2m})$$

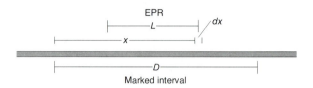

Case III of $D > L$ is the mirror-image of case I.
Adding cases I, II, and III gives

$$R_{D>L} = kL\left[1 - \frac{1}{2m}(1 - e^{-2m}) + \frac{1}{2}\left(\frac{D}{L} - 1\right)(1 - e^{-2m})\right] \quad \text{(A-1b)}$$

Figure A-1 graphs R/kl versus D/L for three values of m. The functions are applicable only to small R values since we have neglected the possibility of two or more EPRs impinging on the marked interval.

A glance at the curves reveals the LNI resulting from the occurrence of more than one exchange in some of the EPRs. The curves bend downward, whereas R would be almost proportional to D for small R values if exchanges were independent. We can visualize how S varies with R by calculating S from Equation 1-6 for the special case of $R_{AB} = R_{BC}$:

$$S = \frac{2R_1 - R_2}{2R_1{}^2} \quad \text{(A-2)}$$

where R_1 is the recombination frequency at any given distance and R_2 is the recombination frequency at twice that distance. Figure A-2 graphs $S(kL)$ versus $2R_1/kL$ for the functions in Figure A-1. (Note: these curves for R and S were calculated on a programmable hand calculator. The curves for S are especially tedious to do without such aid, since, in order to preserve precision in the numerator, $2R_1 - R_2$, they require the handling of R values that have not been rounded. By the same token, the *experimental* assessment of S by two-factor crosses with very close markers is scarcely possible.)

What do the curves tell us and how might we use them? Most important, they "confirm" the contention that exchanges randomly distributed within EPRs (of fixed length and random location) will lead to increasing S as R decreases. What values of S will result? That depends on the value of kL that characterizes the system; kL is just the probability that any site will be covered by an EPR. The maximum value of S is seen to be the reciprocal of

Appendix: Mapping Functions for Very Close Markers

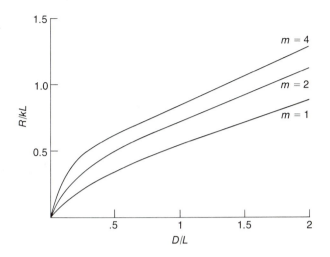

FIGURE A-1

Mapping functions (Equation A-1a, b) for EPRs of fixed length L. The abcissa gives the distance between marked sites measured in multiples of L. The ordinate is the recombination frequency divided by kL, the probability that a particular site will be "covered" by an EPR. The value m is the mean number of Poisson-distributed exchanges per EPR.

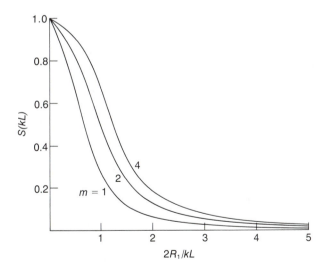

FIGURE A-2

The coefficients of coincidence as a function of recombination frequency for the mapping functions in Figure A-1.

kL independently of m, the mean number of exchanges per EPR. What magnitude are these values? Let us take a reasonable value for kL, such as 0.001 (EPRs one gene long and occurring once per thousand genes). Then, for $m = 2$, when $2R_1 = 10^{-4}$, $S = 1 \times 10^3$; when $2R_1 = 10^{-3}$, $S = 5 \times 10^2$; and when $2R_1 = 10^{-2}$, $S = 5$; S falls to 1 when $2R_1$ is about 0.02. The functions graphed give $S < 1$ for even larger values of R. This is because we have made no allowance for more than one EPR; we have, in effect, supposed complete interference between EPRs. That was a computational convenience.

EPRs with an Exponential Length Distribution

An exponential distribution of lengths is a plausible assumption for EPRs.[3] It results from supposing that "effective pairing" once begun at some point extends until it is terminated by an "accident," which is called that because its probability of occurrence is independent of the length reached by the EPR. This rule leads to a distribution in which the most probable length is the shortest (e.g., one base pair), and increasing lengths are decreasingly probable, with the probability decreasing exponentially with length. For a mean length L, the probability of length x is

$$P_x = ke^{-x/L}\,dx$$

where $e^{-x/L}$ is the probability that the EPR has extended as far as x, and $k(dx)$ is the probability that it suffers its accident right there.

For P_x to be a probability, $\int_{x=0}^{\infty} P_x$ must be unity. That restriction sets $k = 1/L$, and we write

$$P_x = \frac{1}{L} e^{-x/L}\,dx \tag{A-3}$$

The distribution is graphed for several values of L in Figure A-3. The notable feature of these distributions is that with increasing L, large values of x become more probable, but the most probable value always remains the shortest one. We shall employ the same length distribution when we consider splatches and excision tracts. When we do so, we are assuming "accidental" termination of splatch formation or exonucleolytic digestion. The symbols D and m have the same meaning as in our discussion of EPRs of fixed length; L is the mean length of EPRs.

This model makes the supposition that exchanges within EPRs occur at a density that is constant and, in particular, independent of the length of the

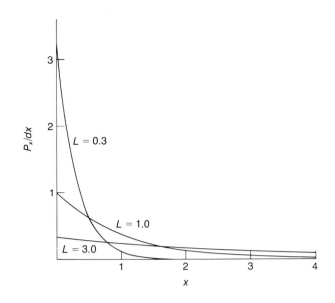

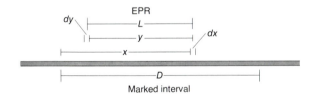

FIGURE A-3
The exponential probability distribution
(Equation A-3) for several values of the
mean, L.

EPR within which they occur. Thus, even when m is large, the shortest EPRs
may have no exchanges at all, while with small m the longest EPRs will have
a great many exchanges. Four cases must be considered.

CASE I.   Both ends of the EPR fall between the marked sites.

The right end of the EPR falls in the interval dx at a distance x from the left
marked site with probability $k(dx)$. With probability $e^{-y/L}$ the EPR will
extend leftward a distance $\geq y$, and with probability $(1/L)dy$ it will end at
the distance y. Recombination will occur if there is an odd number of exchanges in that EPR of length y. Thus,

$$R_1 = \frac{k}{2L} \int_{x=0}^{D} \int_{y=0}^{x} e^{-y/L}(1 - e^{-2my/L})\, dy\, dx$$

$$= \frac{kL}{2} \left[\frac{D}{L} - \frac{D/L}{2m+1} - (1 - e^{-D/L}) \right.$$

$$\left. + \frac{1}{(2m+1)^2}(1 - e^{-(2m+1)D/L}) \right]$$

CASE II.   Only the left end of the EPR falls between the marked sites.

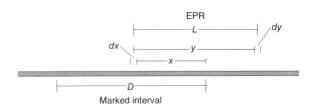

$$R_{II} = \frac{k}{2L} \int_{x=0}^{D} \int_{y=x}^{\infty} e^{-y/L}(1 - e^{-2mx/L}) \, dy \, dx$$

$$= \frac{kL}{2} \left[(1 - e^{-D/L}) - \frac{1}{2m + 1}(1 - e^{-(2m+1)D/L}) \right]$$

CASE III.   Only the right end of the EPR falls between the marked sites; $R_{III} = R_{II}$.

CASE IV.   Neither end of the EPR falls between the marked sites.

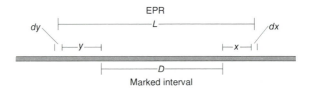

$$R_{IV} = \frac{1}{2}(1 - e^{-2mD/L}) \frac{ke^{-D/L}}{L} \int_{x=0}^{\infty} \int_{y=0}^{\infty} e^{-y/L}e^{-x/L} \, dy \, dx$$

$$= \frac{kL}{2}(e^{-D/L} - e^{-(2m+1)D/L})$$

Adding cases I–IV gives

$$R = \frac{kLm}{2m + 1} \left[\frac{D}{L} + \frac{2m}{2m + 1}(1 - e^{-(2m+1)D/L}) \right] \qquad \text{(A-4)}$$

Figure A-4 graphs R/kl versus D/L for three values of m, and Figure A-5 graphs $S(kL)$ versus $2R_1/kL$ for the same m values. The curves look much like those in Figures A-1 and A-2, respectively. Thus, the assumptions regarding the length distribution of the EPRs appear to make little difference.

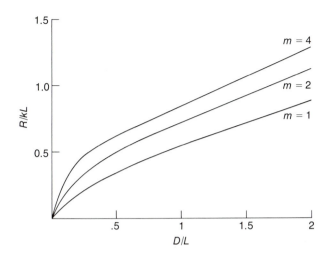

FIGURE A-4
Mapping functions for EPRs with an exponential length distribution (Equation A-4). The mean length of the EPRs is L, and m is the mean number of exchanges in those EPRs of length L.

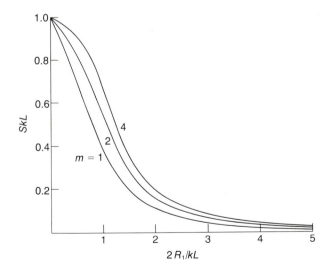

FIGURE A-5
The coefficients of coincidence as a function of recombination frequency for the mapping functions in Figure A-4.

SPLATCHES

In the EPR models we ignored the duplex nature of DNA, or, if you prefer, we supposed that all splatches were of zero length. In this section we develop splatch models and confront the duplex. Our approach remains much the same, but we have to take care to distinguish chromosomes recombinant on only one chain from those recombinant on both. We shall keep our books by weighting the first class by the factor $\frac{1}{2}$. This is fully justifiable for fungi that enjoy a postmeiotic mitosis before nuclei are scored, or for those phages that usually replicate prior to chromosome encapsidation. Again we consider two versions: splatches of fixed length and splatches with an exponential length distribution. In both versions we shall vary the ratio of splices to patches.

SPLATCHES OF FIXED LENGTH[4]

We can categorize splatches in two dimensions: (1) whether they are splices or patches, and (2) according to whether they cover one marked site only, the other marked site only, neither marked site, or both marked sites. Disallowing mismatch correction, splatches that cover both marked sites yield no recombinants. That leaves six interesting cases from the cross in Figure A-6. All involve formation of $a_1^+ a_2^+$ on at least one chain except for the type 3 patch. We draw them all because they are handy when we derive mapping functions for splatches with an exponential length distribution. Without being a bore, it is not easy to be certain that all cases are being counted and properly weighted. For instance, in Figure A-6 the amount of heavy-line DNA on the lower chain is always greater than that on the upper. Clearly, I could draw another six pictures in which the reverse was true. Also, 12 pictures could be drawn to keep track of splatches that are of interest in the formation of the $a_1 a_2$ recombinant. We shall refer, however, only to the cases in Figure A-6; they appear to cover all cases of interest in the correct proportions.

Let k be the probability that the right end of a splatch falls in a small interval dx. Then kD is the probability that it falls in the marked interval. For markers separated by a distance D and splatches of fixed length L, we consider separately the cases of $D < L$ and $D > L$.

For $D < L$, type 1 splices and patches yield recombinants by virtue of the right end of the splatch falling between the sites. If splices are a fraction a of all splatches, then for type 1 splices we write

$$R_{(1s)} = \frac{a}{2} kD$$

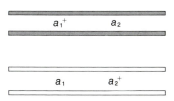

Splices   Patches

| a_1 a_2^+ | | a_1 a_2^+ |
| a_1^+ a_2^+ | Type 1 | a_1^+ a_2^+ |

| a_1^+ a_2^+ | | a_1^+ a_2^+ |
| a_1^+ a_2 | Type 2 | a_1^+ a_2 |

| a_1^+ a_2^+ | | a_1 a_2^+ |
| a_1^+ a_2^+ | Type 3 | a_1 a_2^+ |

FIGURE A-6
Types of splatches (for reference in deriving mapping functions).

where the factor $\frac{1}{2}$ accounts for the fact that only half the molecule is re-combinant. Similarly for type 1 patches, we write

$$R_{(1p)} = \frac{(1 - a)}{2} kD$$

Analogous expressions can be written for type 2 splatches, in which the left end of the splatch falls in the interval. Summing the contributions from type 1 and 2 splatches gives

$$R_{D<L} = kD \qquad\qquad (A\text{-}5a)$$

For the case of $D > L$ we must consider all types except for patches of type 3. Type 1 and 2 splatches each contribute half a recombinant molecule, while type 3 splice contributes a whole recombinant. The weighted contributions for each type are the following:

Splices, type 1 = type 2: $R_{(1s)} = R_{(2s)} = \dfrac{akL}{2}$

Patches, type 1 = type 2: $R_{(1p)} = R_{(2p)} = (1 - a)\dfrac{kL}{2}$

Splice, type 3: $R_{(3s)} = ak(D - L)$

Adding the five types gives

$$R_{D>L} = kL\left[1 - a\left(1 - \frac{D}{L}\right)\right] \qquad \text{(A-5b)}$$

In Figure A-7, R/kL is graphed against D/L for several values of a. We can see LNI in the graphs, all of which show a lesser slope for $D/L > 1$ than for $D/L < 1$. The case where patches are more frequent than splices ($a = 0.2$) bends most sharply, showing the greatest LNI. This conclusion is quantitated by graphs of $S(kL)$ versus $2R_1/kL$ for the three values of a (Figure A-8). In that figure we see that for $2R_1/kL < 1$ all three examples give $S = 0$, i.e., complete positive interference. This is a consequence of our assumption that patches cannot be smaller than L and the convenience assumption that two independent splatches do not occur in the same small interval. For each example, S rises sharply to a maximum (the smallest a value has the largest maximum) and then gradually declines with increasing R. We can anticipate that allowing splatches to vary in length may reduce the upsy-downsy character of S as a function of R.

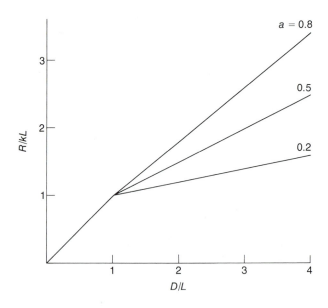

FIGURE A-7
Mapping functions for splatches of fixed length, L (Equations A-5a, b). The fraction of splatches that are splices is a, and kL is the probability that a given site is covered by a splatch.

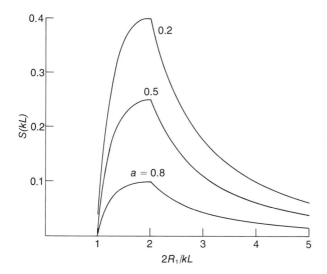

SPLATCHES WITH AN EXPONENTIAL LENGTH DISTRIBUTION

Type 1 splatches (Figure A-6) have their right end in the marked interval, their left end outside the interval, and give half a recombinant. Thus,

$$R_{(1)} = \frac{k}{2} \int_{x=0}^{D} e^{-x/L} \, dx = \frac{kL}{2} (1 - e^{-D/L})$$

Type 2 splatches are mirror-images of type 1, so

$$R_{(1)} = R_{(2)}$$

Type 3 splatches have both ends in the marked interval but only the splices give recombinants. The recombinants produced are whole recombinants, so

$$R_{(3)} = ak \int_{x=0}^{D} (1 - e^{-x/L}) \, dx = akL \left[\frac{D}{L} - (1 - e^{-D/L}) \right]$$

The sum of types 1–3 is[5]

$$R = kL \left[a \frac{D}{L} + (1 - a)(1 - e^{-D/L}) \right] \tag{A-6}$$

In Figure A-9 we graph R/kL versus D/L for three values of a. With this splatch model, R is seen to vary continuously with D, and S (Figure A-10) rises monotonically with decreasing R values.

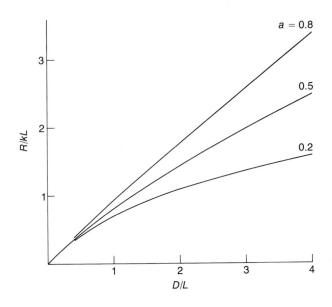

FIGURE A-9
Mapping functions for splatches with an exponential length distribution (Equation A-6). The fraction of splatches that are splices is a; L is the mean splatch length.

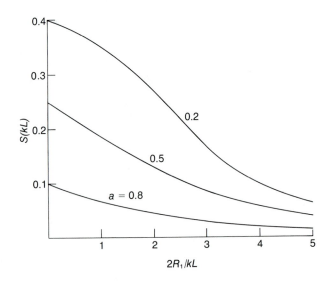

FIGURE A-10
The coefficients of coincidence as a function of recombination frequency for the mapping functions in Figure A-9.

MISMATCH CORRECTION, LNI, AND MAP EXPANSION

LNI and map expansion appear mutually contradictory. LNI requires $R_2 < 2R_1$ for small R, while map expansion is defined by $R_2 > 2R_1$. Nevertheless, mismatch correction can give either, depending on assumptions about its nature, and can even give expansion in one range of R values and LNI in another.

Our models are designed to illustrate effects of mismatch correction isolated from other sources of LNI. Thus, we suppose *splices* of random position (no patches—we do not want their LNI to confuse us). We shall consider three variations within that framework. In each variation we suppose that mismatch correction is equal for all sites and in both directions. That is, the following four reactions occur equally:

$$\frac{a_1^+}{a_1} \rightarrow \frac{a_1^+}{a_1^+}$$

$$\frac{a_1^+}{a_1} \rightarrow \frac{a_1}{a_1}$$

$$\frac{a_2^+}{a_2} \rightarrow \frac{a_2^+}{a_2^+}$$

$$\frac{a_2^+}{a_2} \rightarrow \frac{a_2}{a_2}$$

Excision Tracts of Exponential Length on Splices of Fixed Length

Type 1, 2, and 3 splices as defined in Figure A-6 will produce recombinants without the intervention of correction. Their combined probability is kD. The probability of a splice covering both sites, like this,

is $k(L - D)$, with $D < L$. Mismatch correction can operate on this splice. In order for correction to make a significant contribution to recombination, we must suppose that it is a major route to recombination for very close markers. By that token, the rules governing very close marker recombination guide us in setting conditions for the (hypothetical) operation of mismatch correction. In many cases very close markers do show ever-decreasing recombination frequency with decreasing distance in a more or less additive manner. This suggests we exclude from consideration two possible modes of correction: correction cannot be confined to a point, and it cannot be generally true that excision tracts frequently begin at site *1* and run only left or begin at *2* and run only right. Were either of those situations the rule, then recombination frequency would, with decreasing distance, plateau at a value that was proportional to the correction probabilities of the sites involved. Thus, we shall suppose that excision tracts proceed from one site toward the other. Without making very special assumptions, excision tracts will regularly run from one site toward the other only if each tract is extended in both directions from the heteroduplex site. In writing our mapping functions, however, we need concern ourselves only with that arm of a tract that extends from the initiating site toward the other marked site.

We must calculate $a_1{}^+a_2{}^+$ frequencies resulting from the following two mismatch correction routes (the wavy lines indicate productive excisions):

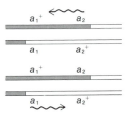

Let ρ be the probability of tract initiation at a heteroduplex site, and l be the mean excision tract length. The various cases will yield either half or full wild-type recombinants with the probabilities indicated:

	Reaction	Product	Probability	Weighting factor
(1)	$\dfrac{a_1{}^+ \quad \overleftarrow{\leavevmode\smash{\sim}}a_2}{a_1 \quad a_2{}^+}$ →	$\dfrac{a_1{}^+ \quad a_2{}^+}{a_1 \quad a_2{}^+}$	$\dfrac{\rho}{2}(1-\rho)(1-e^{-D/l})$	$\dfrac{1}{2}$
(2)	$\dfrac{a_1{}^+ \quad a_2}{\underleftarrow{\smash{\sim}}a_1 \quad a_2{}^+}$ →	$\dfrac{a_1{}^+ \quad a_2}{a_1{}^+ \quad a_2{}^+}$	$\dfrac{\rho}{2}(1-\rho)(1-e^{-D/l})$	$\dfrac{1}{2}$
(3)	$\dfrac{a_1{}^+ \quad \overleftarrow{\smash{\sim}}a_2}{\underleftarrow{\smash{\sim}}a_1 \quad a_2{}^+}$ →	$\dfrac{a_1{}^+ \quad a_2{}^+}{a_1{}^+ \quad a_2{}^+}$	$\dfrac{\rho^2}{4}(1-e^{-D/l})$	1

The probabilities arise from the following considerations.

Reaction 1. $\rho/2$ is the probability of correction at site 2 toward wild type. $1 - \rho$ is the probability of no correction at site 1. $(1 - e^{-D/l})$ is the probability that the excision tract initiated at site 2 fails to reach site 1.

Reaction 2. This is the mirror image of Reaction 1.

Reaction 3. $\rho^2/4$ is the probability of excision toward wild type at both sites. Wild type will indeed result as long as the first tract to be made fails to reach across the marked interval. As before, this probability is $1 - e^{-D/l}$.

Weighting the half recombinants by $\frac{1}{2}$, summing the three cases, and then doubling the sum to include the a_1a_2 recombinant class gives

$$\frac{\rho(2 - \rho)}{2}(1 - e^{-D/l})$$

as the probability that a doubly heterozygous splice will yield recombinants. Recombinants arise without the involvement of correction when splices cover only one marker. For $D < L$, that contribution to R is $kL(D/L)$. The two sources of recombinants add to give

$$R_{D<L} = kL\left[\frac{D}{L} + \frac{\rho(2 - \rho)}{2}\left(1 - \frac{D}{L}\right)\left(1 - e^{-\frac{D}{L}/\left(\frac{l}{L}\right)}\right)\right] \qquad \text{(A-7a)}$$

When $D > L$, correction plays no role. Borrowing from Equation A-5b, and setting $a = 1$,

$$R_{D>L} = kL\left(\frac{D}{L}\right) \qquad \text{(A-7b)}$$

In Figure A-11 R/kL is plotted versus D/L for parameter values that emphasize the contribution from correction ($\rho = 1$ and $l/L = 0.3$) and for a set that deemphasizes it ($\rho = 0.2$ and $l/L = 1$).

In Figure A-12 the plots of S versus R show ever increasing LNI as markers become closer. The values of S drop to zero as the distance between the markers becomes greater than the length of the splice, because of our assumptions of no patches and only one splice. In the next model we eliminate the unaesthetic discontinuity in this one by allowing excision tracts with an exponential length distribution to operate on splices that also have an exponential length distribution.

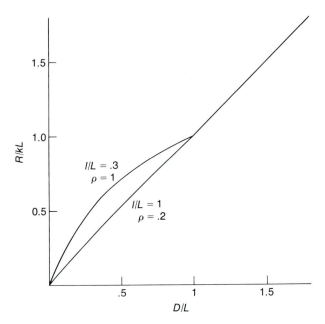

FIGURE A-11
Mapping functions for splices of fixed length L, subject to mismatch correction with excision tracts whose exponential length distribution has mean l (Equations A-7a, b). The probability that a mismatched site will be corrected is ρ.

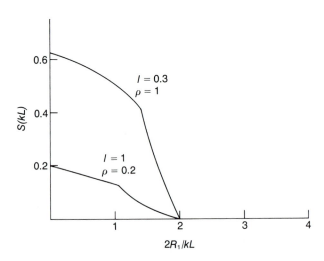

FIGURE A-12
Coefficients of coincidence as a function of recombination frequency for the mapping functions in Figure A-11.

Excision Tracts of Exponential Length on Splices of Exponential Length

To allow for splices having an exponential length distribution, we need only modify the above argument for splices of fixed length. From Equation A-6 we salvage $R = kL(D/L)$ when a, the fraction of splatches that are splices, is unity. This is the contribution we must expect from splices of types 1–3 that cover one marked site only or fall entirely between the marked sites. There will be a fraction of splices, however, that covers both sites, and in this class correction can contribute to R. The fraction of splices covering both sites is $kLe^{-D/L}$, and correction will operate on these to yield recombinants with the same probability as for splices of fixed length. So we can write

$$R = kL\left[\frac{D}{L} + \frac{\rho(2 - \rho)}{2}\, e^{-\frac{D}{L}}\left(1 - e^{-\frac{D}{L}\big/\frac{l}{L}}\right)\right] \tag{A-8}$$

This mapping function is graphed for $l/L = 0.3$ and $\rho = 1$ in Figure A-13, and S is graphed in Figure A-14. As we anticipated, by making the splice lengths exponentially distributed we got the kinks out of the curves. We shall now see that giving the excision tracts a fixed length has more striking consequences.

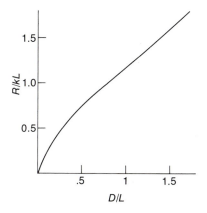

FIGURE A-13
Mapping function for splices of exponential length distribution subject to excision tracts with an exponential length distribution (Equation A-8). The mean splice length is L. The ratio of the mean tract length to the mean splice length, l/L, was set at 0.3, and ρ, the correction probability, was set at unity.

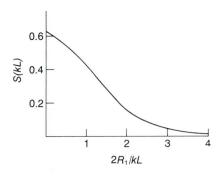

FIGURE A-14
Coefficient of coincidence as a function of
recombination frequency for the mapping
function in Figure A-13.

EXCISION TRACTS OF FIXED LENGTH
ON SPLICES OF FIXED LENGTH

Borrowing pieces from previous arguments, we can write

$$R_{D>L} = kL\left(\frac{D}{L}\right)$$

$$R_{l>D<L} = kL\left(\frac{D}{L}\right) \qquad (A-9)$$

$$R_{l<D<L} = kL\left[\frac{D}{L} + \frac{\rho(2-\rho)}{2}\left(1 - \frac{D}{L}\right)\right]$$

In Figure A-15 we see how this mapping function looks when tracts are rather short ($l = 0.3L$) and correction probability is high ($\rho = 1$). The discontinuities in the curve are, of course, a result of assuming fixed-length splices and fixed-length excision tracts. No matter, they serve to illustrate that mismatch correction *can* give map expansion. Expansion occurs around $D/L = l/L = 0.3$. This is manifested as a rapid rise in R in Figure A-15 and negative values for S in Figure A-16. Note that at larger distances the mismatch correction results in LNI rather than expansion, as revealed by large $S(kL)$ values for $2R_1/kL > 1.3$. As before, $S = 0$ for the shortest and longest distances because of our assumption that two independent splices simply do not occur in very small intervals.

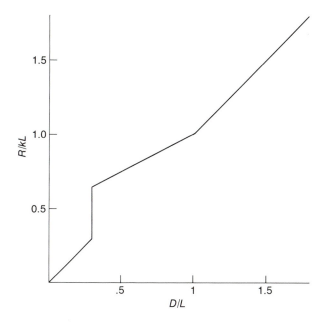

Figure A-15
Mapping function for splices of fixed length, L, operated upon by excision tracts of fixed length (Equation A-9). The probability of correction, ρ, was set at unity, and the excision tract length was set at $0.3L$.

The formalized models illustrate how LNI might result from several factors, i.e., clustering of exchanges, patches, and mismatch correction. Our indicator of LNI, S, has usually been measured in three-factor crosses, since deviations from additivity for very close markers in two-factor crosses are slight even when LNI is strong. For models involving only exchange-clustering and splatches the analysis we have conducted here is (I trust) valid for three-factor crosses, since the models allow no marker effects. For models involving mismatch correction, however, the model based on two-factor crosses cannot describe accurately the three-factor case. In particular, the addition of a third factor is expected to increase the S values above those derived from the two-factor models by adding another source of excision tracts whose two ends serve as nearby points of chain exchange.

We may wonder about the degree of realism in the assumptions of EPRs or splatches of fixed or exponentially varying lengths and the assumption that they are randomly distributed. Each of the assumptions regarding

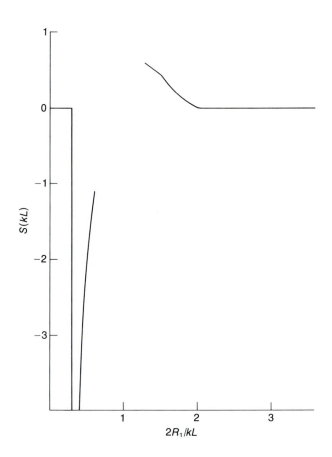

Figure A-16
Coefficient of coincidence as a function of recombination
frequency for the mapping function in Figure A-15.

length distribution is somewhat unreasonable. The assumption of fixed-
length structures (EPRs or splatches) calls for a measuring device without
proposing what that device might be. The assumption of exponentiality
avoids that nicely; if the EPRs or splatches are extended until an accident
terminates them, they will be exponentially distributed. However, that
assumption may be unreasonable for very short structures. In an exponential
distribution the most frequent class is comprised of the shortest structures
(Figure A-3). Depending on molecular mechanisms, EPRs and splatches
that are extremely short might not persist long enough for recombination
to be effected. In fact, it is better that they do not, since recombination might
thereby lose its requirement for homology (Problem 5-3).

I see no problem with the assumption of an exponential length distribution for excision tracts; it seems reasonable from a chemical viewpoint. The alternative assumption of fixed-length (or quasi-fixed length) tracts is plausible but requires an assumption of either a measuring device or a sequence-dependent probability of tract termination.

The mathematically convenient assumption that EPRs and splatches are randomly positioned is contrary to the evidence, discussed in Chapter 6, that suggests they are fixed in position. However, the assumption does not make the mapping functions totally worthless. Assuming a random location for EPRs and splatches is at least approximately equivalent to assuming that the markers used to measure R are disposed at random with respect to the EPRs or splatches. This will be more or less true in many cases. Anyway, the models are primarily meant to illustrate certain notions; it seems unlikely that they will be critical in experimental discrimination of mechanisms.

Crosses that record the segregation of flanking markers as the position and distance separating two very close markers is varied may help us to understand the nature of the apparently fixed elements marking one or both of the boundaries of an EPR or a splatch. In the following sections we formalize the principal assumptions underlying several fixed-element models.

FIXED PAIRING SEGMENTS

Consider two marked sites within the same fixed pairing segment. Add flanking markers like this:

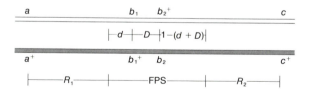

Here D is the distance between marked sites 1 and 2, and d is the distance of the marked interval from the left end of the FPS. Both are measured as fractions of the length of the FPS. R_1 and R_2 are the recombination frequencies observed in the crosses $ab_1 \times a^+b_1^+$ and $b_2^+c \times b_2c^+$, respectively. The distances corresponding to R_1 and R_2 are large compared to L, so that R_1 can be considered to depend on the distance between A and the left end of the FPS involving B while R_2 depends on the distance from C to the right end of that FPS.

POISSON-DISTRIBUTED EXCHANGES WITHIN FPS
WITH TOTAL INTERFERENCE ON NEARBY PAIRINGS

In its most primitive version, pairing between homologous FPSs prevents pairing of nearby FPSs, so that among $b_1{}^+b_2{}^+$ recombinants all genotypes other than a^+c are a result of multiple exchanges within the FPS. These exchanges are supposed to be Poisson-distributed in number and randomly located. Haldane's Equation describes the probability of an odd number of exchanges in any segment of the FPS once assurance of pairing is obtained. Formation of $b_1{}^+b_2{}^+$ provides that assurance. Let x be the mean number of exchanges per pairing of the FPS. Then the frequencies of the genotypes for flanking markers among $b_1{}^+b_2{}^+$ can be written

$$R_{ac^+}^+ = \tfrac{1}{4}(1 - e^{-2dx})(1 - e^{-2(1-d-D)x})$$

$$R_{a^+c}^+ = \tfrac{1}{4}(1 + e^{-2dx})(1 + e^{-2(1-d-D)x})$$

$$R_{ac}^+ = \tfrac{1}{4}(1 - e^{-2dx})(1 + e^{-2(1-d-D)x}) \qquad \text{(A-10)}$$

$$R_{a^+c^+}^+ = \tfrac{1}{4}(1 + e^{-2dx})(1 - e^{-2(1-d-D)x})$$

With three parameters, d, D, and x, how shall we proceed? Let us consider the following cases.

1. Locate the *1–2* interval centrally on the FPS. Then vary D at a fixed value of x and vary x at a fixed value of D. The frequencies of the genotypes for flanking markers are graphed in Figures A-17 and A-18, respectively.

2. Let $D = 0$ and move the marker pair across the FPS from left to right; i.e., let d vary from 0 to 1. Genotype frequencies for this case are graphed in Figure A-19, for both $x = 0.5$ and $x = 2$.

Figure A-17 demonstrates that a small marked interval, centrally placed, gives comparable frequencies for all four types of flanking marker combinations even with a mean number of exchanges per FPS as low as 2. As the interval is expanded, all types except a^+c, the single recombinant type, vanish.

Figure A-18 reveals that reducing the mean number of exchanges (x) per FPS has the same consequences as increasing the length of the marked interval (D).

Figure A-19 reveals coordinate variation in the frequencies of the types as the location of the marked interval within the FPS is varied. The terminal positions ($d = 0$ and $d = 1$), which maximize the differences between the recombinant types, a^+c and ac^+, also give the largest differences between the parental types, a^+c^+ and ac. As the marked interval is moved from one end of the FPS to the other, the parental classes, but not the recombinants, reverse their relative frequencies.

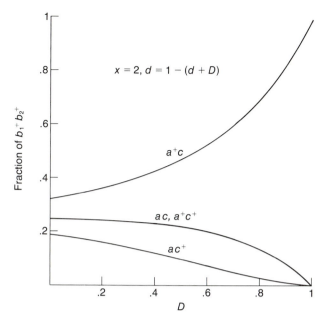

FIGURE A-17

Genotype frequencies of flanking markers as functions of the distance, D, between two close markers in a simple FPS model (Equation A-10). The cross is:

$$
\begin{array}{ccc}
a & b_1 b_2{}^+ & c \\
\hline
& \times & \\
a^+ & b_1{}^+ b_2 & c^+ \\
\hline
\end{array}
$$

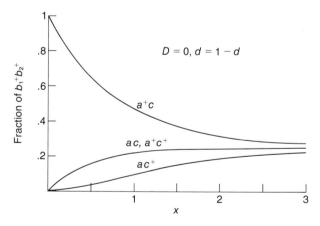

FIGURE A-18

Genotype frequencies of flanking markers as functions of the number of exchanges, x, per pairing segment in a simple FPS model (Equation A-10). The cross is $a b_1 b_2{}^+ c \times a^+ b_1{}^+ b_2 c^+$, as in Figure A-17.

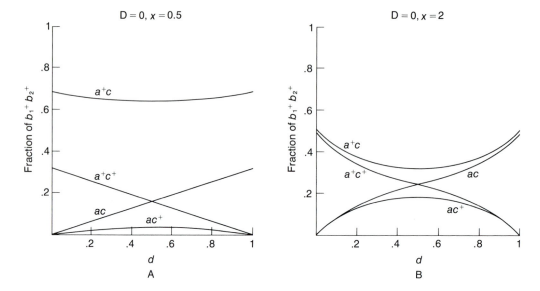

Genotype frequencies of flanking markers as functions of the position of close markers within pairing segments in a simple FPS model (Equation A-10). The cross is $ab_1b_2{}^+c \times a^+b_1{}^+b_2c^+$

FPSs HAVE ONE OR TWO EXCHANGES, AND ONE-EXCHANGE FPSs INTERFERE WITH NEARBY PAIRING

Parentals for flanking markers arise whenever there is a double exchange in the FPS with only one of the two exchanges falling in the marked interval. Which parental type is obtained will depend on whether the other exchange occurs left or right of the interval. Of the parental types so created, some will be lost due to independent exchange in the AB and BC intervals. Then, if p is the probability of one exchange,

$$ac = 2d \ (1 - p)(1 - R_1 - R_2)/[p + 2(1 - p)(1 - D)] \qquad \text{(A-11a)}$$

$$a^+c^+ = 2(1 - p)(1 - d - D)(1 - R_1 - R_2)/[p + 2(1 - p)(1 - D)] \qquad \text{(A-11b)}$$

The more frequent recombinant type for flanking markers, a^+c, arises in one of two ways: in one step as a single exchange falling between sites 1 and 2, or in two steps with a double exchange in which one element only falls between sites *1* and *2* accompanied by an independent exchange in the intervals *AB* or *BC*.

$$a^+c = \{p + 2(1 - p)[dR_1 + (1 - d - D)R_2]\}/[p + 2(1 - p)(1 - D)] \tag{A-11c}$$

The rare recombinant type arises only by a double exchange in the FPS followed by an independent exchange on one side or the other.

$$ac^+ = 2(1 - p)[dR_2 + (1 - d - D)R_1]/[p + 2(1 - p)(1 - D)] \tag{A-11d}$$

In most crosses of this type, R_1 and R_2 are chosen to be safely outside a supposed FPS but close enough that exchanges independent of the ones producing the $b_1{}^+b_2{}^+$ recombinant are not too frequent. For illustration, let us settle on $R_1 = 0.2$ and $R_2 = 0.1$, varying p, D, *and* d to see how the model behaves.

Pick a centrally located interval and see how the genotype frequencies (among $b_1{}^+b_2{}^+$) change as D expands from zero to 1. The curves for $p = \frac{1}{3}$ and $p = \frac{2}{3}$ are shown in Figure A-20. The curves for $p = \frac{1}{3}$ look like those in

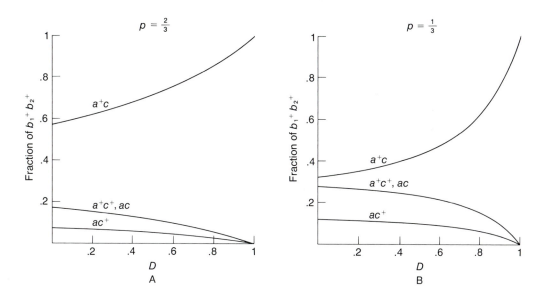

FIGURE A-20

Genotype frequencies of flanking markers as functions of the distance, D, between close markers in a modified FPS model (Equation A-11 with $R_1 = 0.2$ and $R_2 = 0.1$). p is the fraction of FPSs with one exchange. The cross is $ab_1b_2{}^+c \times a^+b_1{}^+b_2c^+$.

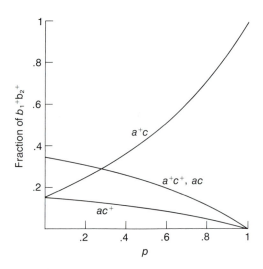

FIGURE A-21
Genotype frequencies of flanking markers as functions of the fraction of FPSs with one exchange, p, in a modified FPS model (Equation A-11 with $R_1 = 0.2$ and $R_2 = 0.1$). D was set at 0 and d at 0.5. The cross is $ab_1b_2{}^+c \times a^+b_1{}^+b_2c^+$.

Figure A-17, illustrating the experimental difficulties encountered when one tries to choose among the models.

Let p vary from 0 to 1 for a centrally located interval of length $D = 0$ (Figure A-21). When $p = 0$, all exchanges in the FPS are doubles, so the two parental types, a^+c^+ and ac, predominate. The two other types are a result of superimposed exchanges in the intervals AB and BC. When $p = 1$, all exchanges in FPS are singles, and the resulting interference prevents exchanges in the flanking intervals; hence, only the recombinant a^+c is formed.

Hold p constant (at $\frac{1}{3}$ and $\frac{2}{3}$) and D constant at zero. Then move the marked interval across the FPS by varying d from zero to one. The resulting genotype frequences are graphed in Figure A-22. The parental types vary much as they do for the primitive FPS model graphed in Figure A-19. The recombinant types, however, vary with d only because I chose the two flanking intervals unequal in size. Since inequality of these intervals is experimentally common, however, variation in recombinant frequencies will typically accompany variation in parental frequencies as markers are chosen in different parts of an FPS.

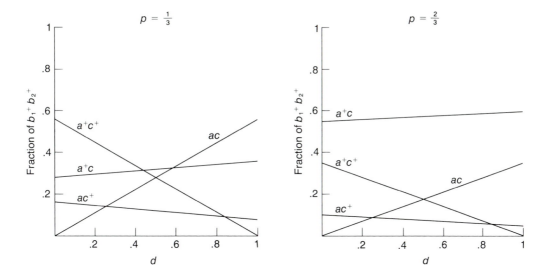

FIGURE A-22

Genotype frequencies of flanking markers as functions of the position of close markers within an FPS (Equation A-11 with $R_1 = 0.2$ and $R_2 = 0.1$). p is the fraction of FPS with one exchange. The cross is $ab_1b_2{}^+c \times a^+b_1{}^+b_2c^+$.

FIXED STARTING POINTS

In this version of the FSP model the chromosome has a large number of starting points for splatch formation with only a small probability of splatch formation per point. In any one act a splatch extends one way or the other from a starting point. Splatches that turn out to be splices impose complete positive interference on the flanking intervals while patches impose none. We let the splatch length distribution be exponential with mean L. We measure distance between adjacent FSPs:

Now consider how the various types of $b_1{}^+b_2{}^+$ recombinants arise in the cross $ab_1b_2{}^+c \times a^+b_1{}^+b_2c^+$, where R_1 and R_2 are large compared to the

R value characterizing the interval from the left to the right FSP. Let the fraction of splatches that are splices be p; then $1 - p$ is the fraction that are patches.

Parental types require patches. They arise when the variable end of a patch falls between sites 1 and 2. Some of these parentals then get "lost" by independent exchanges in the flanking intervals AB and BC. The parental type ac can form when a patch comes from the left FSP, while a^+c^+ requires a patch from the right.

The recombinant type a^+c forms by one of two routes: splices from either side ending in the b_1b_2 interval produce a^+c, or some of the "lost" parental types show up as a^+c. The ac^+ recombinant arises only via "lost" parental types. Then

$$ac = (1 - p)e^{-d/L}(1 - R_1 - R_2)/(e^{-d/L} + e^{-(1-d-D)/L})$$
$$a^+c^+ = (1 - p)e^{-(1-d-D)/L}(1 - R_1 - R_2)/(e^{-d/L} + e^{-(1-d-D)/L})$$
$$a^+c = p + [(1 - p)(R_1e^{-d/L} + R_2e^{-(1-d-D)/L})]/(e^{-d/L} + e^{-(1-d-D)/L})$$
$$ac^+ = (1 - p)(R_2e^{-d/L} + R_1e^{-(1-d-D)/L})/(e^{-d/L} + e^{-(1-d-D)/L})$$

$$(A\text{-}12)$$

Let us examine these expressions. First, put the marked interval halfway between the left and right FSPs and expand it from $D = 0$ to $D = 1$. At $d = 1 - (d + D)$ the equations reduce to

$$a^+c^+ = ac = (1 - p)(1 - R_1 - R_2)/2$$
$$a^+c = p + [(1 - p)(R_1 + R_2)]/2 \qquad (A\text{-}13)$$
$$ac^+ = (1 - p)(R_1 + R_2)/2$$

These frequencies are independent of the length of the interval, depending only on the fraction of splatches that are splices and on the frequencies of independent events in the flanking intervals. We can graph them versus p to see how the frequencies of the types vary as there are relatively more or less splices to patches (Figure A-23). Not surprisingly, the graphs are reminiscent of those in Figure A-21, in which FPSs enjoyed one or two exchanges with interference imposed only by FPSs with one exchange. How do the genotype frequencies change as a small interval is moved from the left to the right FSP?

The expressions in Equation A-12 (at $D \to 0$) are graphed versus d for two different values each of p and L. When patches predominate ($p = 0.1$),

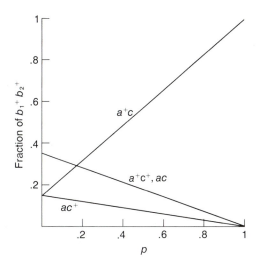

FIGURE A-23
Genotype frequencies of flanking markers as functions of the fraction, p, of splatches that are splices in an FSP model (Equation A-13 with $R_1 = 0.2$ and $R_2 = 0.1$). The cross is $ab_1b_2{}^+c \times a^+b_1{}^+b_2c^+$.

parentals (ac and a^+c^+) outnumber recombinants. In this situation the recombinants arise in large part via parentals "lost" by exchange in the flanking intervals, and the single recombinant need not be much more frequent than the triple (Figures A-24A,C). When patches and splices are equally frequent ($p = 0.5$), recombinants outnumber parentals because of parentals "lost" by exchanges in the flanking intervals (Figures A-24B,D). Furthermore, the more frequent recombinant is in great excess of the less frequent one.

Comparison of Figures A-22B and A-24D reveals the most distinctive difference between the FPS and FSP models: the models are essentially mirror images of each other with regard to the genotype frequencies they predict. The FPS model predicts that the parental type whose right-hand marker (c in Figure A-22B) is like that of the more frequent recombinant type (a^+c) will increase in frequency as b_1b_2 move rightward. The FSP model (Figure A-24D) predicts that the parental type whose right-hand marker (c^+) is like that of the *less* frequent recombinant (ac^+) will increase as b_1b_2 move rightward.[7]

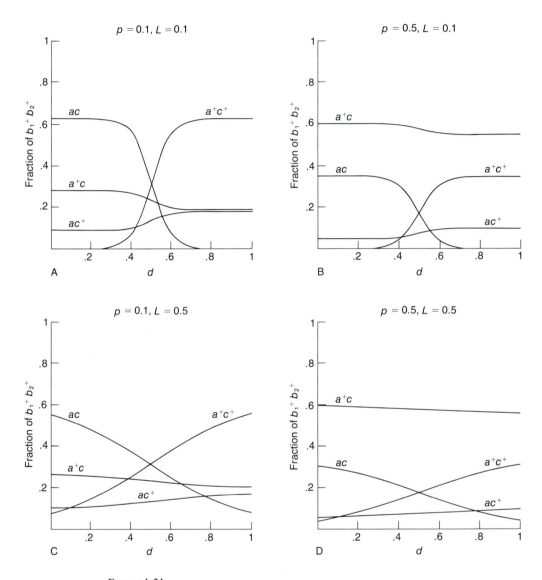

FIGURE A-24
Genotype frequencies for flanking markers as functions of the position of two close markers between two FSPs (Equation A-12 with $R_1 = 0.2$, $R_2 = 0.1$, and $D \to 0$). The fraction of splatches that are splices, p, and the mean splatch length, L, are as follows:

	a	b	c	d
p	0.1	0.5	0.1	0.5
L	0.1	0.1	0.5	0.5

The cross is $ab_1b_2{}^+c \times a^+b_1{}^+b_2c^+$.

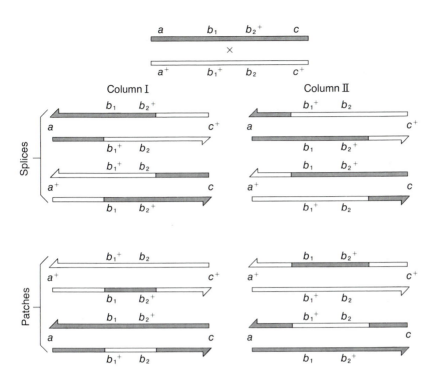

FIGURE A-25
All possible splatches covering sites b_1 and b_2.

MISMATCH CORRECTION AND FSPs

"Polarized" recombination* might result from chemically polarized excision tracts. In our discussion of mismatch correction operating on uniformly located splices (pp. 261–262) we ruled out the possibility of unidirectional excision because its frequent occurrence would lead to a distance-independent recombination frequency for markers so close that they were generally embraced by the same splatch. However, the mean length of excision in the two chemical directions could be different, and if only certain splatches are allowed, polarity can result.[8] Consider the cross shown in Figure A-25. Eight splatches simultaneously involving sites 1 and 2 can arise. The splices in column I arise, we presume, from an FSP on one side of gene B, while those

---

* I use "polarized" to mean $ac \neq a^+c^+$ among $b_1{}^+b_2{}^+$ recombinants when $R_1 = R_2$.

in column II arise from an FSP on the other. The patches in the two columns are distinguished not by the side on which they originate but by the chain that is patched.

Now, let T be the mean of an exponential distribution of excision tract length and $t < T$ be the mean tract length in the opposite chemical direction. We will arbitrarily assign the two tract lengths to the two chain directions this way:

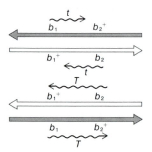

Then, the eight kinds of splatches have correction potentials like those shown in Figure A-26.

SPLATCHES FROM COLUMN I ONLY

This case illustrates the origins of the polarity. Of the two patches in column I, the lower one will produce more $b_1{}^+b_2{}^+$ than will the upper; in the lower patch many tracts end between the marked sites, while in the upper patch fewer tracts end between the sites. The result is that ac will be more frequent among $b_1{}^+b_2{}^+$ than will a^+c^+. (Had we opted to deal with splatches in column II, the outcome would have been reversed.) in order to isolate polarity due to excision in this case and subsequent ones, we suppose that any splatch that covers one site covers the other site as well.

We borrow from page 263 to write the frequencies of the four flanking marker types among $b_1{}^+b_2{}^+$ recombinants. The frequency of $b_1{}^+b_2{}^+$ that are ac^+ is

$$R_{ac^+}^+ = \frac{p}{2} \frac{\rho(2 - \rho)}{4} (1 - e^{-D/t})/R^+$$

where D is the distance from site 1 to site 2, p is the fraction of splatches that are splices, ρ is the correction probability, and R^+ is the total $b_1{}^+b_2{}^+$

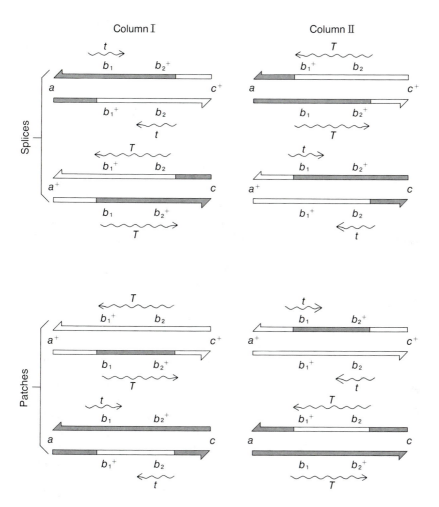

Figure A-26
Correction potentials to B^+ for double-site splatches when the mean lengths (t and T) of excision tracts depend on chain polarity.

frequency. The analogous expressions for the other types are

$$R_{a^+c}^+ = \frac{p}{2}\frac{\rho(2-\rho)}{4}(1 - e^{-D/T})/R^+$$

$$R_{a^+c^+}^+ = \left(\frac{1-p}{2}\right)\frac{\rho(2-\rho)}{4}(1 - e^{-D/T})/R^+$$

$$R_{ac}^+ = \left(\frac{1-p}{2}\right)\frac{\rho(2-\rho)}{4}(1 - e^{-D/t})/R^+$$

These frequencies reduce to expressions that are independent of ρ, the correction probability:

$$\begin{aligned}
R_{ac^+}^+ &= p(1 - e^{-D/t})/F \\
R_{a^+c}^+ &= p(1 - e^{-D/T})/F \\
R_{a^+c^+}^+ &= (1 - p)(1 - e^{-D/T})/F \\
R_{ac}^+ &= (1 - p)(1 - e^{-D/t})/F
\end{aligned} \tag{A-14}$$

where

$$F = (1 - e^{-D/T}) + (1 - e^{-D/t})$$

To get some feeling for these expressions, we shall evaluate them at $p = 0.3$, $D/T = 0.2$, and $D/t = 5$. I get

$$R_{ac^+}^+ = 0.254$$

$$R_{a^+c}^+ = 0.046$$

$$R_{ac}^+ = 0.592$$

$$R_{a^+c^+}^+ = 0.108$$

The polarity is exemplified by $R_{ac}^+ > R_{a^+c^+}^+$. An inequality between the recombinant types ac^+ and a^+c is expected; in a simple EPR or splatch model that inequality signals the order of sites *1* and *2* with respect to the flanking markers. In these, and most, models the rare type arises by the more complex event, e.g., a triple exchange in the EPR case. Note that the inequality calculated, however, signals the wrong order! Had we chosen to allow splatches from column II only rather than column I, the observed inequality for the recombinant, as well as the parental flanking marker types, would have been reversed. The site order deduced from the recombinant types would then have been correct. Thus, the model divorces site order from flanking marker assortment and warns us of the possible need to secure independent evidence of order when trying to disentangle very close marker recombination phenomena.

SPLICES FROM BOTH COLUMNS BUT PATCHES FROM COLUMN I ONLY

In this case suppose that recombination in gene B is under the influence of FSPs to its left and right. The left FSP initiates splices and patches like those in column I; the right FSP initiates splices like those in column II but patches like those in column I. Let d be the distance from the marked sites to the left FSP and $1 - d$ the distance to the right FSP. Let the splatches have an exponential distribution of lengths of mean L. We retain the assumption that splatches that cover one site always cover the other as well, so that all $b_1{}^+b_2{}^+$ arise by the correction process being examined in this model.

We now recognize that ac^+, for example, among $b_1{}^+b_2{}^+$ can arise by one of two routes: a splatch from the left, subject to short, effective excision tracts, and a splatch from the right, subject to long, relatively ineffective excision. The two routes must be weighted according to the distance of sites 1 and 2 from the left and right FSPs, respectively. This gives

$$R_{ac^+}^+ = p[(1 - e^{-D/t})e^{-d/L} + (1 - e^{-D/T})e^{-(1-d)/L}]/G$$
$$R_{a^+c}^+ = p[(1 - e^{-D/T})e^{-d/L} + (1 - e^{-D/t})e^{-(1-d)/L}]/G$$
$$R_{a^+c^+}^+ = (1 - p)(1 - e^{-D/T})/F$$
$$R_{ac}^+ = (1 - p)(1 - e^{-D/t})/F$$

(A-15)

where L is the mean splatch length, F is defined in Equation A-14, and $G = F(e^{-d/L} + e^{-(1-d)/L})$. The parent types, a^+c^+ and ac, are independent of d as they were in A-14. The recombinant types, however, do depend on d. For the special case of $L \gg 1$, $R_{ac^+}^+ = R_{a^+c}^+$ and both are independent of d. Figure A-27 graphs the four R^+ values for $p = 0.3$, $D/T = 0.2$, $D/t = 5$, and $L - 0.5$. The "polarity" of the model is the inequality in R_{ac}^+ and $R_{a^+c^+}^+$. Note that the two recombinant frequencies signal the wrong site order when the markers are near the left FSP ($d < 0.5$) and the correct one when they are near the right site. They give no signal when $d = 0.5$.

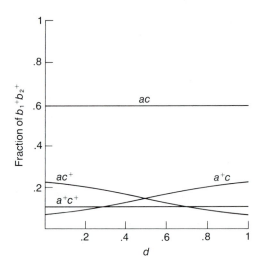

FIGURE A-27
Genotype frequencies of flanking markers as functions of the position of two close markers between FSPs that initiate splatches on the same chain subject to polarized excision tracts (Equation A-15, with $p = 0.3$, $D/T = 0.2$, $D/t = 5$, and $L = 0.5$). The cross is $ab_1b_2^+c \times a^+b_1^+b_2c^+$.

SPLICES AND PATCHES FROM BOTH COLUMNS

In this case splices and patches arising at the left FSP are the kind in column I; splatches arising at the right FSP are those in column II. We borrow from Equation A-15 and get

$$R_{ac^+}^+ = p[(1 - e^{-D/t})e^{-d/L} + (1 - e^{-D/T})e^{-(1-d)/L}]/G$$

$$R_{a^+c}^+ = p[(1 - e^{-D/T})e^{-d/L} + (1 - e^{-D/t})e^{-(1-d)/L}]/G$$

$$R_{a^+c^+}^+ = (1 - p)[(1 - e^{-D/T})e^{-d/L} + (1 - e^{-D/t})e^{-(1-d)/L}]/G \qquad \text{(A-16)}$$

$$R_{ac}^+ = (1 - p)[(1 - e^{-D/t})e^{-d/L} + (1 - e^{-D/T})e^{-(1-d)/L}]/G$$

These R^+ are graphed in Figure A-28 with the same parameter values used in Figure A-27.

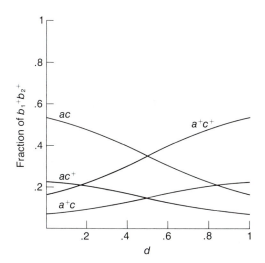

FIGURE A-28
Genotype frequencies of flanking markers as functions of the position of two close markers between FSPs that initiate splatches on opposite chains subject to polarized excision tracts (Equation A-16 with parameter values as in Figure A-27). The cross is $ab_1b_2^+c \times a^+b_1^+b_2c^+$.

LNI FROM A POLARIZED EXCISION MODEL

We can write a mapping function and calculate S as a function of R for a special case of the polarized excision model. As in previous models, we let the splatch that covers one site always cover the other as well. This assumption isolates LNI due to excision. With this assumption we need not distinguish between splices and patches and we can write

$$R = \frac{kL}{2}\left[\frac{\rho(2-\rho)}{2}(1-e^{-D/T}) + \frac{\rho(2-\rho)}{2}(1-e^{-D/t})\right]$$

$$= \frac{kL\rho(2-\rho)}{4}\left[2 - e^{-D/T} - e^{-D/t}\right]$$

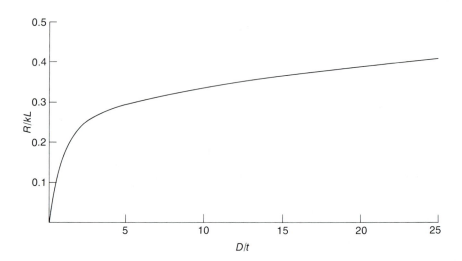

FIGURE A-29
Mapping function resulting from polarized excision (Equation A-17 with $\rho = 1$ and $T = 25t$).

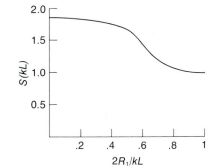

FIGURE A-30
The coefficient of coincidence as a function of recombination frequency for the mapping function in Figure A-29.

In Figure A-29 R/kL is graphed versus D/t for $\rho = 1$ and $T = 25t$. The curve has two limbs, corresponding to production of wild types by short tracts when D is small and by both long and short tracts when D is large. The graph implies a point we made earlier. If $t = 0$, i.e., if the short tracts in our model had zero length, then recombination frequencies due to correction would not extrapolate to zero with decreasing distance. Instead, they would reach a minimum equal to the extrapolation back to zero of the second limb of the curve.

The dependence of S on R is graphed in Figure A-30.

PROBLEMS

A-1 (a) Write a mapping function for very close markers assuming fixed-length excision tracts and fixed-length splatches, half of which are splices and half patches. Graph R/kL versus D/L for your function and compare your graph with one in the literature (J. R. S. Fincham and R. Holliday (1970) *MGG* **109**: 309).

     (b) Examine your graph in part *a*. Show the following:

        (i)   When the length (l) of excision tracts is zero, recombination frequency fails to extrapolate to zero as distance goes to zero.

        (ii) When the length (l) of excision tracts is greater than the length (L) of splatches, mismatch correction makes no contribution to recombination and your model becomes identical to the splatch model in which lengths are fixed and half the splatches are splices (Figure A-6 at $a = 0.5$).

     (c) Graph $S(kL)$ versus $2R_1/kL$ for your function. On the same graph indicate $S(kL)$ for the fixed-length splatch model in which half the splatches are splices (Figure A-7 at $a = 0.5$).

A-2 (a) Write a mapping function for excision tracts of exponential length distribution operating on splatches of exponential length distribution with patches and splices equally frequent. Graph your mapping function. Compare it with the noncorrection model of exponential splatches (Equation A-6 and Figure A-8 evaluated at $a = 0.5$).

     (b) Graph $S(kL)$ vs $2R_1/kL$. Compare the graph with the corresponding model lacking mismatch correction (Figure A-9 evaluated at $a = 0.5$).

A-3 Identify the distinctive features of each model in the appendix and propose an experiment (original or otherwise) to test for the presence of that feature in natural recombination.

A-4 Locate the fallacy in the following discussion of a model for fungal recombination (F. W. Stahl [1969] *G* **61** [Suppl.]:1.)

The nonreciprocal exchanges involve the formation of overlap regions [i.e., splices] as in phage. When an overlap falls upon a marker, a postmeiotic segregation ... results. Such overlaps will show up as "map expansion" ... in linkage data. For very close markers only one of the two duplex chains will be recombinant. For more distant markers, both will be. Postmeiotic mitosis will yield only one wild-type spore for very close markers while yielding two for more distant ones.

NOTES

1. R. H. Pritchard (1960) *GR* **1**:1.
2. F. W. Stahl, R. S. Edgar, and J. Steinberg (1964) *G* **50**:539.
3. Ibid.
4. J. R. S. Fincham and R. Holliday (1970) *MGG* **109**:309.
5. This equation is related to one in H. Bernstein (1962) *JTB* **3**:335.
6. K. M. Fisher and H. Bernstein (1965) *G* **52**:1127.
7. R. Holliday and H. L. K. Whitehouse (1970) *MGG* **107**:85.
8. J. R. S. Fincham (1976) *H* **36**:81.

Solutions to Problems

CHAPTER 1

1-1 (a) Parental types are a^+b and ab^+; recombinant types are a^+b^+ and ab.

   (b) The parental types equal each other; the recombinant types equal each other.

   (c) Each of the recombinant types is less than or equal to 25 percent in frequency since the recombination frequency is never observed to exceed 50 percent.

1-2 (a) The frequency of a^+b^+ is 30 percent, so ab must be 30 percent, while a^+b and ab^+ are each 20 percent.

   (b) a^+b and ab^+ are each 30 percent, a^+b^+ and ab are each 20 percent.

1-3 (a) The order is

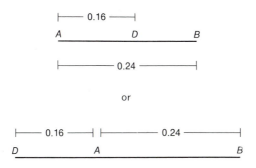

or

   (b) The order is DAB.

1-4 Equation 1-6 states that $S = \dfrac{R_{DA} + R_{AB} - R_{DB}}{2R_{DA}R_{AB}} = \dfrac{0.16 + 0.24 - 0.38}{2(0.16)(0.24)} = 0.26.$

1-5 (a) $S = \dfrac{R_{AB} + R_{BC} - R_{AC}}{2R_{AB}R_{BC}} = \dfrac{0.1 + 0.1 - 0.25}{2(0.1)(0.1)} = -2.5.$

The negative value of S tells us that the marker b influences R_{AB} and/or R_{BC}, so their sum is less than R_{AC}.

(b) $S = R_{\text{double}}/R_{AB}R_{BC} = 0.005/(0.1)(0.1) = 0.5.$
$R_{AC} = R_{AB} + R_{BC} - 2R_{\text{doubles}} = 0.19.$

In this cross the presence of b in one parent apparently lowers R_{AC} to 0.19 from the value 0.25 observed in its absence. Two hypotheses should be considered. (1) Heterozygosity at B disrupts the exchange process in the interval A-C as well as in A-B and B-C. For example, if B were an inversion it might have such an effect. (2) The marker b is a deletion. In b chromosomes the interval A-C is physically shorter than when the chromosomes are b^+.

(c) When both parents are b, the smaller value for R_{AC} is obtained. This result tells us that the state of heterozygosity at b is not essential for the reduction in R_{AC}, while it is fully compatible with the hypothesis that b is a deletion. Can you suggest other mechanisms for the role of b in reducing recombination frequency in the A-C interval?

1-6 (a) The most frequent complementary genotypes, ab^+c and a^+bc^+, must be the parental types.

(b) $R_{AB} = 0.09 + 0.01 = 0.10;\ R_{BC} = 0.09 + 0.19 = 0.28;$
$R_{AC} = 0.19 + 0.01 = 0.20.$

(c)

(d) $b^+ac \times ba^+c^+$ yields b^+a^+c and bac^+ as double recombinants.

(e) $S = R_{\text{double}}/R_{BA}R_{AC} = 0.01/(0.10)(0.20) = 0.5.$

1-7 An unrectified map is

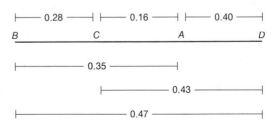

Since $S = 1$ at all R values, the Haldane Function (Figure 1-5) will rectify the map, transforming the R values to distance (x). The resulting rectified map looks like this:

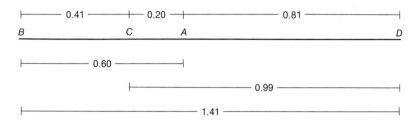

1-8–10 The data in Problems 1-8 and 1-9 are solutions to the equation (Equation 1-7) graphed in Figure S-1. Starting with the Haldane Function (Equation 1-2), I multiplied the exponent by the factor $2x/(1 - e^{-2x})$ to inject interference into the function. The Haldane Function is shown for comparison. Your points for

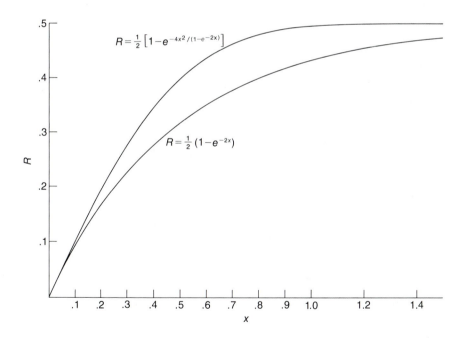

FIGURE S-1
A concocted mapping function illustrating interference (*upper curve*) compared with the Haldane Function (*lower curve*), which describes recombination frequency (R) versus distance (x) in the absence of interference.

$R = f(x)$ in Problems 1-8 and 1-9 will fall along the curve for Equation 1-7, if neither you nor I have made mistakes.

CHAPTER 2

2-1 (a) On the average, three particles are adsorbed per cell. The particles will be approximately Poisson-distributed among cells so that $P_O = e^{-3} = 0.050$ is the fraction of cells with no particles.

(b) $P = e^{-6} = 0.0025$.

(c) Fraction of mixedly infected cells is $(1 - e^{-3})^2 = 0.90$.

(d) $e^{-10} = 4.5 \times 10^{-5}$; $e^{-20} = 2.1 \times 10^{-9}$; $(1 - e^{-10})^2 = 0.99991$.

2-2 When multiplicity is 6, g is 0.80. When multiplicity is 20, g is about 0.95.

2-3 If matings are fast, they will not interfere with one another, and we may suppose them to be Poisson-distributed among lineages. Then, for an average number of matings, $m = 3$, the fraction of lineages with no matings is $e^{-m} = e^{-3} = 0.050$. For $f = \frac{1}{2}$ (and infinite input), half of the matings could lead to color conversion, so the mean number of such matings is $\frac{3}{2} = 1.5$. The fraction of phages that had no such matings is $e^{-1.5} = 0.22$.

2-4 Substituting in Equation 2-2 gives $r = \frac{1}{2}(1 - e^{-3(0.2)}) = 0.23$.

2-5 Recombinant frequencies depend on the relative frequencies of the two infecting types by the factor $f(1 - f)$, where f is the fraction of one type and $(1 - f)$ the fraction of the other. For the cross performed, $f = 5/(5 + 15) = 0.25$, and for the cross desired, $f = 0.5$. Thus, had Jan not goofed he would have obtained $r = 0.082(0.25/0.1875) = 0.11$.

2-6 In Figure 2-3 we see that the finite-input factor, g, for a multiplicity of infection of 4 is 0.66. When multiplicity is 14, g is 0.92. Thus, had Gus's stocks been high titer he would have observed $r = 0.2(0.92/0.66) = 0.28$.

2-7 For large mR, Equation 2-2 becomes $r = g/2$. When the multiplicity of infection is 15, $r = 0.93/2 = 0.47$.

2-8 (a) The data indicate the following gene order: $CDABEGF$.

(b) The maximum r value is 0.31 judging from the crosses $C \times E$, $C \times G$, $C \times F$.

(c) When r has reached a maximum, we may suppose that $R = 0.5$, so that in Equation 2-2, $r = \frac{1}{2}(1 - e^{-0.5m})$. Solving for m at $r = 0.31$ gives $m = 1.94$ (call it 2).

(d) To rectify the map, we convert the r values in part (a) to additive distances. First convert r to R using Equation 2-2, then convert those R values to x with the Haldane Equation (Equation 1-2). Then draw the map to scale so that the distances between the genes are proportional to those x values. My map is shown in Figure S-2. Did the function work? Are the distances in fact additive? They look good to me, except for widely separated markers. But that's OK; when r is so large as to be essentially independent

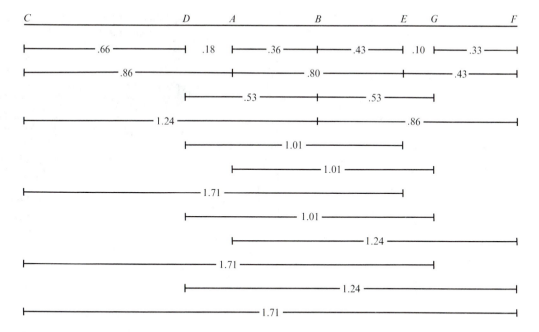

FIGURE S-2
Genetic linkage map for Problem 2-8d.

of distance, we cannot expect distances calculated from those large r values to be accurate.

2-10 (a) $r = (g/2)(1 - e^{-m/2})$

(b) $r = (0.9/2)(1 - e^{-m_1/2}) = 0.32; m_1 = 2.5$

(c) $\bar{r}/r_1 = \dfrac{2gf(1 - f)d(\bar{m} + 1)}{2gf(1 - f)d(m_1 + 1)} = \dfrac{\bar{m} + 1}{m_1 + 1} = 2; \bar{m} = 6.0$

(d) $m_2 = 9.5$

CHAPTER 3

3-1 (a) In each tetrad two of the four cells were type a and two were a^+; two were b and two were b^+.

(b) Accompanying the a^+b^+ spore must have been the spore of complementary genotype, ab. The other two spores must have been a^+b and ab^+. The tetrad is tetratype (T).

(c) ab^+, ab^+, a^+b, a^+b, is a PD tetrad.

(d) a^+b^+, a^+b^+, ab, ab, is a NPD tetrad.

3-2 (a) All tetrads will contain $2a$ and $2a^+$ spores without regard to distance of A from its centromere.

(b) When both A and B are remote from their respective contromeres, they will be distributed at meiosis independently of their centromeres and therefore completely at random with respect to each other. The distribution of two b^+ and two b at random with respect to the two a^+ and two a will result two-thirds of the time in a T tetrad. The same result will obtain as long as either A or B is remote from its centromere.

If the genetic distance of each gene from its centromere is zero, then the frequency of T tetrads will be zero. If the distance of one gene from its centromere is zero, and if the other is x units from its centromere (where x is a small distance), then the frequency of T tetrads will be about $2x$, since half the members of a T tetrad are recombinant. If both are close, one being x units and the other y units from its respective centromere, then $T = 2(x + y)$.

(c) Cross one of the strains marked in a gene tightly linked to its centromere with the strain bearing the third marker. A small T frequency indicates that the third marker is close to its centromere.

3-3 (a) Inseparable linkage results in all tetrads being PD.

(b) When R_{AB} is small, those bivalents in which an exchange occurs will have only one exchange and will all be T. Thus, $R_{AB} = 0.01$ implies $T = 0.02$, since only half the spores in a T tetrad are recombinant.

(c) This is an occasion to point out that $R = T/2 + \text{NPD}$. Therefore, when $R = 0.40$ and $T = 0.70$, $\text{NPD} = 0.05$. The small value of NPD tells us that double exchanges are rather rare, so let us enumerate them and the singles but ignore higher numbers of exchanges. In the absence of chromatid interference, $\text{NPD} = 0.05$ implies that $4 \times 0.05(= 0.20)$ of the bivalents had two exchanges (Figure 3-2). Half of the 0.20 resulted in T tetrads, so that of the total 0.70 Ts, 0.60 have one exchange and 0.10 have two. Then $x = 0.60/2 + 0.20 = 0.50$. Compare this answer with the mapping function characterized by interference that you worked out in Problems 1-8, 1-9, and 1-10.

3-4 In the absence of chromatid interference, one-fourth of the double exchange tetrads are PD, half are T, and one-fourth are NPD for genes A and C.

3-5 (a) Exactly one exchange in each bivalent between A and B yields all T (see Figure 3-2).

(b) With exactly two exchanges, $\text{PD} = \text{NPD} = \frac{1}{4}$, $T = \frac{1}{2}$ (see Figure 3-2).

(c) With a third exchange, all the PD and NPD from part b become T, and half of the T remain T, the rest becoming equally PD and NPD. That gives $\text{PD} = \text{NPD} = \frac{1}{8}$, $T = \frac{3}{4}$.

(d) With an infinite number of exchanges, the markers in genes A and B assort at random, giving $\text{PD} = \text{NPD} = \frac{1}{6}$, $T = \frac{2}{3}$.

(e) We need $T = f(n)$ that oscillates $1, \frac{1}{2}, \frac{3}{4}$ with $n = 1, 2, 3$ and steadies out at $\frac{2}{3}$ as n approaches infinity. $T = \frac{2}{3}[1 - (-\frac{1}{2})^n]$ does the job.

3-6 This is a bit of algebra. We must weight each $T = f(n)$ by the probability of n, assuming that n is Poisson-distributed with mean X. Then add up all the terms.

$$T = \sum_{n=0}^{\infty} P_n f(n) = \sum_{n=0}^{\infty} \frac{X^n e^{-X}}{n!} \frac{2}{3}\left[1 - \left(-\frac{1}{2}\right)^n\right]$$

which gives

$$T = \frac{2}{3}(1 - e^{-3X/2})$$

3-7 The answer is shown in Figure S-3.

3-8 See the dot in Figure S-3. It corresponds to $T = 0.7$ at $x = 0.5$ and $X = 1.0$.

3-9 (a) Exchanges of hot-hot or cold-cold chromatids do not result in isolabeling. These are expected to be half the total. Thus, one-half of the bivalents are expected to show terminal isolabeling, and did.

(b) If no bivalent has more than one chiasma, then we expect $S = 0$ for all values of R (but beware of complications with very close markers, described in later chapters).

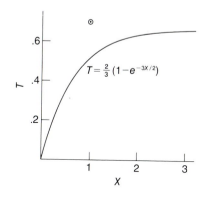

FIGURE S-3
The frequency of tetratype tetrads (T) as a function of the number of crossovers per tetrad (X) (between nonsister strands) in the absence of chromatid and chiasma interference.

CHAPTER 4

4-1 (a) Half the cases of redundancy for gene A will be heterozygous. Thus, the redundancy frequency for gene A is 0.02. Assuming that all genes are equally likely to be redundant, the length of the redundancy is 0.02 of the length of the chromosome, or 4×10^3 base pairs.

(b) $2(0.1)(0.9)(0.02) = 0.0036$.

4-2 (a) The redundancy length in the deletion phages will be the normal length plus the length of the deletion $= 0.02 + 0.04 = 0.06$. Gene A will therefore be redundant with probability 0.06 and heterozygous with probability 0.03.

(b) It will be normal, i.e., 1 percent.

4-3 (a) Peak 1 contains a^+r phages; 2 contains $a^+\,r^+$; 3 contains ar; and 4 contains ar^+.

(b) Since exchanges occur at random with respect to genotype, the number of phages of intermediate density will be half of those that have had an exchange. Since peaks 2 and 3 contain about half the progeny particles, we can conclude that all the particles have had an Int-mediated exchange.

(c) $1:2:3:4 = 1:2:2:4$.

4-4 (a) I count 46 bursts with hc^+mi^+ phages and 50 bursts with the complementary type h^+cmi. Since there were 81 total bursts, the probability of finding both types in a burst if their presence is uncorrelated is $(50/81) \times (46/81) = 0.351$, and the number of bursts expected with both types is $81 \times 0.351 = 28.4$. Similarly, the number of bursts expected with only hc^+mi^+ is $46(81 - 50)/81 = 17.6$, and the number expected with only $h^+cmi = 50(81 - 46)/81 = 21.6$. These expectations compare with the observed numbers as follows:

	Observed	Expected
Both types	46	28.4
hc^+mi^+ only	0	17.6
h^+cmi only	4	21.6
Neither	31	13.4
Total	81	81.0

(b) For the two noncomplementary classes hc^+mi^+ and $hcmi^+$, the observations and expectations are as follows:

	Observed	Expected
Both types	26	25.0
hc^+mi^+ only	20	21.0
$hcmi^+$ only	18	19.0
Neither	17	16.0
Total	81	81.0

Similar calculations for the other pairs of classes support the conclusion that the noncomplementary types are at most weakly correlated, while the complementary types are strongly correlated.

4-5 The procedure is the same as described for Problem 4-4.

	Observed	Expected
Both	38	34.8
hc^+mi^+ only	6	9.2
h^+cmi only	15	18.2
Neither	8	4.8
Total	67	67.0

	Observed	Expected
Both	39	36.1
hc^+mi^+	5	7.9
$hcmi^+$	16	18.9
Neither	7	4.1
Total	67	67.0

For both complementary and noncomplementary types, the classes of "both" and "neither" are in excess of expectation, indicating a weak positive correlation in the occurrence of all recombinant types, complementary or noncomplementary.

4-6 With $g = 0.4$, the mapping function for λ becomes $r = 0.2[1 - e^{-m(1-e^{-2x})/2}]$. For essentially unlinked markers, i.e., markers for which $R \cong 0.5$, we write $r = 0.2(1 - e^{-m/2}) = 0.15$, from which we calculate $m = 2.8$. In Figure S-4

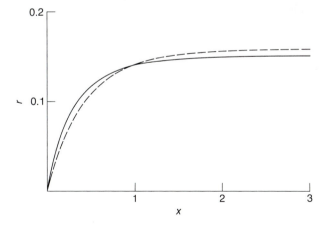

FIGURE S-4
A mapping function for λ in which $g = 0.4$ and $m = 2.8$ (*solid curve*) and $g = 0.8$ and $m = 1.0$ (*dashed curve*).

the function is plotted and compared with that for $g = 0.8$ and $m = 1.0$. The impression that the two functions would be hard to distinguish with mapping data is supported by a few calculations of s. I find s falling between 3 and 4, which is close to the plateau values reported (Figure 2-5) and the values we calculated in Chapter 2 when $g = 0.8$ and $m = 1.0$. Thus, the properties of the λ map can be accounted for as well with high cell-to-cell heterogeneity and several rounds of mating as they can by lower heterogeneity and fewer matings.

We can set a lower limit to g as follows. For the function $r = g/2[1 - e^{-m(1 - e^{-2x})/2}]$, let m approach infinity, which reduces the function to $r = g/2$. Since r values as large as 0.15 are observed, g cannot be less than 0.3. This g value corresponds to an effective multiplicity of about two.

There is a special case that permits a direct determination of g in λ. In *red gam* crosses dimerization via a Rec-mediated exchange is prerequisite to encapsidation. If g were unity, this requirement would lead us to expect $r = 0.5$ for markers at opposite ends of the virion chromosome. However, the r values observed are only about 0.2 (J. M. Crasemann, personal communication). Thus, for this case, g has an effective value of 0.4. However, the considerations that suggested small g values cannot be pertinent to these crosses, in which replication continues in the theta mode throughout. The small g value may be due instead to a tendency to intraclonal mating due to proximity, perhaps between sisters during or after replication.

CHAPTER 5

5-1 (a) $(L - D)/L$.

(b) $(L - D/L = 0.3; L = D/0.7$.

(c) 50 amino acids $\equiv$ 150 base pairs. $L = 150/0.7 = 214$ base pairs.

5-2 The probability that a splatch known to cover site 1 will reach at least to site 2 is $P = e^{-D/L}$, where L is the mean splatch length. For $P = 0.3$ and $D = 150$ base pairs, $L = 125$ base pairs. Exponential length distributions are explained more fully in the appendix.

5-3 (a) The chance that any given sequence of two bases, reading from left to right, is a particular sequence is $(\frac{1}{4})^2$. The number of times the particular sequence would occur in a chromosome 10^7 base pairs long is about $(\frac{1}{4})^2 \times 2 \times 10^7 = 1 \times 10^6$. The factor of two recognizes that the sequence in question can occur in the duplex with a left-right or a right-left orientation.

(b) A given sequence of three base pairs will occur about $(\frac{1}{4})^3 \times 2 \times 10^7 = 3 \times 10^5$.

(c) We may suppose that sequences that are long enough to be rare are Poisson-distributed among chromosomes many times their length. If a sequence is known to occur once, the probability that it will not occur again in that chromosome is e^{-m}, where m is the expected number of occurrences. For $e^{-m} = 0.99$, $m = 0.01$, and we can write $(\frac{1}{4})^N \times 2 \times 10^7 = 0.01$, for which $N = 16$. That appears to be at least an approximate value for the length a sequence must be if we are to be 99 percent certain that it does not repeat in the chromosome. Of course there are many sequences of this length in the chromosome, about 10^7, and the probability is essentially unity that some of the 16-mers will repeat. So how long must sequences be if we are to be 99 percent certain of *no* repeats in a chromosome 10^7 base pairs long? That will be found as the solution for N in the expression $(\frac{1}{4})^N \times 2 \times 10^7 \times 10^7 = 0.01$, for which $N = 27$. I am not satisfied that the calculations above are rigorous, but I think they are decent approximations. They tell us that a splatch-forming mechanism must verify a minimum of about three twists of a helix before committing itself.

5-4 For illustration, suppose wild type is AT. Then the transition mutant must be GC, and two types of heteroduplexes can arise. They are GT and AC. These two types are chemically distinct and therefore distinguishable, at least in principle. If we could trace the origins of the DNA segments of the duplexes, we could distinguish subclasses of these two types of heteroduplexes. Labeled parent molecules

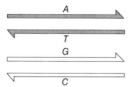

allow us to distinguish four different kinds of patches:

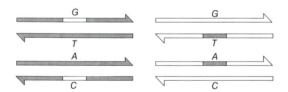

and four kinds of splices:

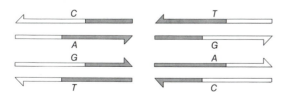

In the next chapter we shall encounter experiments in λ that successfully distinguish among several of these types.

CHAPTER 6

6-1   We double the wild-type frequencies to obtain the recombinant frequencies:

Cross	R
$a_1a_2{}^+ \times a_1{}^+a_2$	0.02
$a_2a_3{}^+ \times a_2{}^+a_3$	0.01
$a_1a_3{}^+ \times a_1{}^+a_3$	0.026

Then $S = \dfrac{r_{12} + r_{23} - r_{13}}{2r_{12}r_{23}} = \dfrac{0.02 + 0.01 - 0.026}{2(0.02)(0.01)} = 10.$

6-2   (a) R_{doubles} = twice the frequency of $a^+b^+c^+ = 4 \times 10^{-5}$. Then

$$S = \frac{R_{\text{doubles}}}{R_{ab} \times R_{ac}} = \frac{4.0 \times 10^{-5}}{(1.0 \times 10^{-3})(2.0 \times 10^{-3})} = 20$$

(b) $R_{ac} = R_{ab} + R_{bc} - 2SR_{ab}R_{bc} = 1.0 \times 10^{-3} + 2.0 \times 10^{-3} - 8 \times 10^{-5}$
$= 2.92 \times 10^{-3}$.

(c)   If S is 40, $R_{ac} = 2.84 \times 10^{-3}$. Most measurements of recombination frequency would not distinguish 2.8×10^{-3} from 2.9×10^{-3}. Thus, especially for small R values, S is not easily determined from two-factor crosses alone.

6-3   The problem commands us to let $D = 1$ when $R = 0.001$. So we calculate:

$$R_2 = 2R_1 - 2SR_1{}^2$$
$$= 2(0.001) - 2(100)(0.001)^2$$
$$= 0.0018$$

Then we calculate:

$$R_4 = 2R_2 - 2SR_2{}^2$$
$$= 2(0.0018) - 2(99)(0.0018)^2$$
$$= 0.0030$$

and finally

$$R_8 = 2R_4 - 2SR_4{}^2$$
$$= 2(0.0030) - 2(83)(0.0030)^2$$
$$= 0.0045$$

Reiterating this procedure gave me the following values:

D	R_i	$2R_i$	S	R_{2i}
1	0.001	0.002	100	0.0018
2	0.0018	0.0036	99	0.0030
4	0.0030	0.0060	83	0.0045
8	0.0045	0.0090	33	0.0076
16	0.0076	0.0152	2.6	0.015
32	0.015	0.030	0.023	0.030
64	0.030	0.060	0.019	0.060
128	0.060	0.12	0.30	0.12
256	0.12	0.24	0.55	0.22
512	0.22	0.44	0.85	0.36
1024	0.36	0.72	0.99	0.46
2048	0.46	0.92	1.0	0.50
4096	0.50	1.0	1.0	0.50

Our tables will be a bit different because of inaccuracies in reading S values from the graph. R is graphed against D in Figure S-5. In D units, a hypothetical FPS appears to have a length of about 7; when $R = 0.002$ ($D = 2$), S starts to plunge, reaching half its maximum value at $R = 0.004$ ($D = 7$). In Figure S-5 the dashed line indicates the recombination frequency if every exchange in the (hypothetical) FPSs were contributing to R. In fact, only FPSs with odd numbers

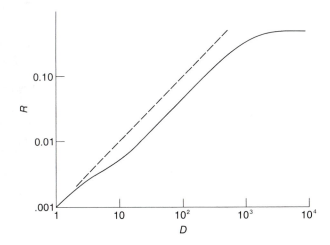

FIGURE S-5
A mapping function constructed by the boot strap. The dashed line is the extrapolate of the initial, linear portion of the function.

of exchanges are being detected when D is greater than about 10. The discrepancy between the straight line and the mapping function allows us to estimate the mean number of exchanges per FPS. Since the discrepancy is about twofold, about half the exchanges are going undetected above $D = 10$. If we make assumptions about the distribution of exchanges among FPSs, we can calculate a mean (m). We shall make two sets of assumptions and calculate two means.

1. Exchanges are Poisson-distributed among FPSs. At smaller D, R is proportional to the mean number of exchanges per FPS. At larger D, R depends on the probability of an odd number of exchanges in an FPS. The twofold discrepancy between the two limbs of the mapping function then leads us to write

$$2 = \frac{m}{\frac{1}{2}(1 - e^{-2m})}$$

for which m is about 0.8.

2. Suppose FPSs have either one or two exchanges, but never more (and never zero). Then, if p is the fraction with 1 exchange,

$$2 = \frac{p + 2(1 - p)}{p}$$

for which $p = \frac{2}{3}$, and $m = 1.3$.

The results of the two calculations agree in estimating a smallish number of exchanges per FPS.

6-4 (a) Let p be the frequency of correction to wild type at site a, q be the frequency at site b, and r be that at site c. We can then write

$$\frac{1}{4}(p + q) = 0.037$$

$$\frac{1}{4}(p + r) = 0.030$$

$$\frac{1}{4}(q + r) = 0.043$$

The three unknowns work out to be $p = 0.05$, $q = 0.1$, and $r = 0.07$.

(b) Let s be the frequency of correction to wild type at site d. Then $\frac{1}{4}(p + s) = 0.033$. Since $p = 0.05$, $s = 0.08$. We can then predict the other two cases:

$$bd^+ \times b^+d \qquad \frac{1}{4}(q + s) = 0.045$$
$$cd^+ \times c^+d \qquad \frac{1}{4}(r + s) = 0.038$$

Our assumptions of independent, low-frequency correction are supported if the predicted wild-type frequencies are realized in experiments.

(c) The discrepancy you note between your calculated prediction and the observation for the DNA $b^+e \times be^+$ suggests that sites b and e are co-corrected, presumably because of their proximity.

6-5 (a) $R_{13} > R_{12} + R_{23}$ is called "map expansion."

(b)
$$S = \frac{R_{13} + R_{23} - R_{13}}{2R_{12}R_{23}} = -2.5 \times 10^3$$

S must be equal to or greater than zero in the absence of marker effect. Therefore, map expansion is a marker effect.

(c) The marker effect causing this map expansion could have any of several causes. (1) The marker at site 2 might be a deletion or some other gross chromosomal change (see Problem 1-5). (2) Two marked sites very close together (like sites 1 and 2 or sites 2 and 3) might block splatch formation more than two sites further apart, such as sites 1 and 3. (3) Recombinants may arise by mismatch correction with a typical tract length longer than the distance between sites 1 and 2 or sites 2 and 3 but shorter than the distance between sites 1 and 3. Did you think of other possible causes for map expansion?

CHAPTER 7

7-2 (a) Since there is equal probability of correcting a site to wild type or mutant type, i.e., $\rho/2$, the various asci will occur with the following frequencies:

$$5^m:3^+ = 5^+:3^m = \rho(1 - \rho)$$
$$6^m:2^+ = 6^+:2^m = \rho^2/4$$
$$4:4 \text{ aberrant} = (1 - \rho)^2$$
$$4:4 \text{ normal} = \rho^2/2$$

When $\rho = 0.4$, the fraction of abnormal asci that are 4:4 aberrant is $(1 - \rho)^2/(1 - \rho^2/2) = 0.39$. At $\rho = 0$ all abnormal asci are aberrant 4:4; at $\rho = 1$ none are.

(b) From such data, it appears that splatches are not formed reciprocally in yeast.

(c) The model calls for half of the aberrant tetrads to be tetratype and half to be parental ditype for flanking markers.

7-3 (a) Aberrant tetrads arise by two routes, splices and patches, in the model. Half arise by splices, becoming tetratypes that will remain so because of interference. The half that arise by patches will have more diverse fates.

Since patches impose no interference in our model, nonsister chromatids may exchange in one or the other of the flanking intervals (but not both, because of interference). R for the flanking markers is $0.10 + 0.05 = 0.15$, so that $2 \times 0.15 = 0.30$ of the tetrads that are patched at the converting site will manifest recombination of flanking markers. Thus, of all aberrant asci, $0.50 + 0.50(0.30) = 0.65$ will be tetratype. The remainder (0.35) will be parental ditype for the flanking markers, since our assumption of strong positive chiasma interference forbids the formation of nonparental ditype tetrads.

(b) According to the model with which we are dealing, half of each type of aberrant ascus is recombined for flanking markers as a result of the splice that led to the observed aberration. In addition, independent exchanges recombine flanking markers in those aberrant asci arising from patches. Of the patched asci, 0.30 will be so recombined, but in only half of these cases will the recombining chromatids be of identical genotype. Thus, 0.575 of the 6:2 asci will contain two chromatids that are recombinant for flanking markers and identical at the converting site. Since 0.65 of the 6:2 asci are tetratype, $0.575/0.65(=0.88)$ of the 6:2 tetratypes will be recombined for flanking markers between two chromatids that have identical genotypes for the converting site.

7-4 Let a be the fraction of splatches that are splices. Then, the fraction of 6:2 asci recombined for flanking markers between chromatids identical at the converting site will be $a + (1 - a)R = 0.5$; with $R = 0.15$, $a = 0.41$.

7-5 There is nothing intrinsically paradoxical about nonreciprocal recombination giving consistent maps.

7-6 Co-conversion can give map expansion only if it is due to mismatch correction (see below) or is subject to other marker effects.

Co-conversion due to mismatch correction will give map expansion only if certain distributions in the lengths of excision tracts obtain. In particular, map expansion will *not* be observed if excision tracts have an exponential length distribution (see the appendix).

CHAPTER 8

8-4. (a) Let Y be the mean number of recombinogenic events (i.e., splices plus patches) per bivalent in the marked interval. Two markers that are not very close together, i.e., whose recombination is primarily by chromosome exchange rather than by conversion, will recombine if either of the following conditions are met:

1. If there is one recombinogenic event, and if that event is accompanied by exchange of flanking material, then half the strands in that bivalent will be recombinant. The contribution to R from this route is therefore $Ye^{-Y}/4$.

2. If there is more than one event in the interval, then half the strands in

that bivalent will be recombinant (assuming no chromatid interference). The contribution to R from this route is $\frac{1}{2}(1 - e^{-Y} - Ye^{-Y})$. Adding the two routes gives the mapping function

$$R = \frac{1}{2}\left(1 - e^{-Y} - \frac{Y}{2}e^{-Y}\right)$$

Since only half of the events exchange flanking material, and since Y counts events per bivalent rather than events per chromatid pair, we must define a new scale in order to compare this function with Haldane's. If we let $Y = 4x$, we get

$$R = \frac{1}{2}(1 - e^{-4x} - 2xe^{-4x})$$

which we can compare with the Haldane Function, $R = \frac{1}{2}(1 - e^{-2x})$.

Both functions give $R = x$ when x is small, and $R = 0.5$ when x is large. They are graphed in Figure S-6.

Our newly derived function is more nearly linear at small values of x, so it does indeed have positive interference. This interference is quantitated in the graph of S versus $2R_1$ in Figure S-7. By the way, do you agree that this mapping function equally well described a creature in which there is no chiasma interference but strong chromatid interference (Chapter 3)? The idea *is* tidy from a formal point of view. A molecular explanation for the proposed behavior is needed to make it totally attractive, however. Try to find some data on S versus R in yeast (*S. cereviseae*) to compare with the graphed predictions of the model.

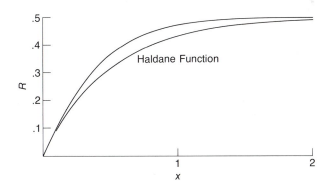

FIGURE S-6
A mapping function for Poisson-distributed splatches in which splices and patches alternate. The splatch length was set at zero for convenience in computing this function for markers that are not very close.

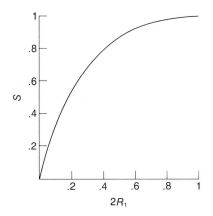

FIGURE S-7
The coefficient of coincidence as a function of recombination frequency for the mapping function in Figure S-6.

(c) For the case in which each third recombinogenic event gives exchange of flanking material, I get

$$R = \frac{1}{2}\left[1 - \left(1 + \frac{2Y}{3} + \frac{Y^2}{6}\right)e^{-Y}\right]$$

With the change of scale to give $R = x$ for close markers, this gives

$$R = \tfrac{1}{2}[1 - (1 + 4x + x^2)e^{-6x}]$$

which indeed shows stronger interference than the function in part *a*.

8-5 Among tetrads convertant at B that have not recombined for A and C, the fraction with recombination in the DC interval will be twice that in the overall population of tetrads. Demonstration of this kind of "negative interference" would constitute strong support for the model; the model would be ruled out by a significant failure to find the "negative interference." Data in one paper (Fogel and Hurst, 1967) do argue against the model, if I have understood that paper correctly. Can you find other relevant data, preferably in yeast?

8-6 In addition to the markers at which aberrant segregation is being monitored, add an unlinked marker of the sort that does not give postmeiotic segregation. For simplicity and maximum sensitivity try to use a marker whose total rate of aberrant segregation is very low. Then some of the 4:4 aberrants for the test locus will be manifest as tetrads in which two of the eight spores have unique genotypes, unlike each other and unlike any other spores in the ascus. If one or the other (or both) of the loci are far from their respective centromeres, then two-thirds of the aberrant 4:4 asci will be detected by this test (see Problem 3-2).

CHAPTER 10

10-3 We can borrow our function from Equation 2-7, which was written for T4. We do not need both terms since we are not supposing our phage (λ?) to have a permuted chromosome. Thus $r = 2gf(1 - f)(1 - e^{-md})$, which for $f = \frac{1}{2}$ becomes $r = g(1 - e^{-md})/2$. Our problem states that r plateaus at 0.15 as d becomes large, so we write $r_{max} = g/2 = 0.15$, which implies $g = 0.3$; our phage is acting as if only about two of the infecting particles are participating (see Figure 2-3). The completed function is $r = 0.15(1 - e^{-md})$. We can compare this with our first mapping function for λ (Figure 2-6) in two ways.

1. What sorts of coefficients of coincidence (s) does the function manifest? If we substitute r in place of R in Equation A-2 and simplify, we get $s = 3.3(1 - e^{-md})$. As long as md is large, $s = 3.3$ (compare with the data in Figure 2-5). As md goes to zero, s goes to zero and the positive interference imposed on individual matings is apparent. However, for small d values LNI (due to patches and mismatch correction, presumably) might swamp the reducing value of s so that "in a not quite trivial sense it may be permissible to say that positive interference is not found because it is obscured by negative interference" (Hershey, 1958).

2. Graph the mapping function so that it coincides at $r_{max}/2 = 0.075$ with the function in Figure 2-6. Note in Figure S-8 that the two curves then lie close to each other throughout. In regraphing the function from Figure 2-6 I have picked $m = 0.94$ (instead of 1.0) to bring r_{max} exactly to .15, and I have adjusted the abscissas of the two functions so that the functions coincide at $r_{max}/2$ as well.

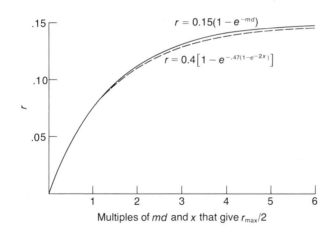

FIGURE S-8
A mapping function for λ that assumes $S = 0$ in individual matings (*solid curve*). The finite-input factor, g, was adjusted to 0.3 to fit the observed maximum recombinant frequency of 0.15. The dashed curve is essentially the original λ mapping function from Chapter 2, which assumes $g = 0.8$ and $S = 1$.

CHAPTER 11

11-1 The nonreciprocal-hybrid-DNA model envisions the following outcomes to a recombinational interaction:

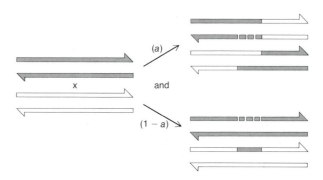

If we give our symbols the same meaning they have in the appendix, then we can borrow from Equation A-5.

CASE I: $D < L$.

1. First consider the contribution to R from splatches, i.e., from chromatids that *do* have hybrid DNA. Since they are half the products, we write $R = kD/2$.

2. To these recombinants we must add those that arise from the chromatid that has its exchange points at the same place on both chains. These chromatids are a fraction $a/2$ of the total, and they are recombinant on both chains, so that $R = akD/2$.

Combining contributions (1) and (2) gives $R_{D<L} = (1 + a)kD/2$ and

$$\frac{R_{D<L}}{kL} = \left(\frac{1 + a}{2}\right)\frac{D}{L} \tag{1}$$

CASE II: $D > L$.

1. Again we first consider the contribution to R from the chromatids with splatches. Recognizing that they are half of all chromatids and borrowing from Equation A-56, we write

$$\frac{R}{kL} = \frac{1}{2}\left[1 - a\left(1 - \frac{D}{L}\right)\right]$$

2. To this term we must again add a term for those recombinants that arise without a splice. That term will be the same as in step 1 in case I. Then combining

terms gives

$$\frac{R_{D>L}}{kL} = \frac{1}{2}(1 - a) + \frac{aD}{L} \qquad (2)$$

(a) Equations 1 and 2 are the mapping function we seek.
(b) The ratio of the final to the initial slope is $2a/(1 + a)$. Only when $a = 1$, i.e., when all splatches are splices, will this ratio be unity. For the case of $a = 0.5$, the ratio is $1/1.5$, rather than 1.0 as claimed.

11-2   Class I models are described by the splatch models in the appendix. Class II models are described by the model you elaborated in Problem 11-1. Can you show that class III models give about the same functions as the class I models (see Bernstein, 1962)? When I let the splices be tiny and the length distribution between paired exchanges be exponential, I get $R = 2kL(D/L + 1 - e^{-D/L})/3$, which, if the $\frac{2}{3}$ is absorbed into the constant kL, is identical to Equation A-6 at $a = \frac{1}{2}$.

11-5   Let L be the length of the sex circle, D the distance between the markers, and k the probability of sex circle formation. In one case, both exchanges occur in the marked interval. This will occur with probability $k(D/L)^2$. Half of the occurrences will result in two recombinant chromatids. Thus, this case will make a contribution to R of $k(D/L)^2$. In the second case, one exchange occurs in the interval and the other outside. This will make a contribution to R of $2k(D/L)(L - D)/L$. Combining the two cases gives the mapping function $R/k = D(2 - D/L)/L$.
    A graph of R/k versus D/L (Figure S-9) reveals downward curvature characteristic of LNI. The LNI is quantitated in a graph of Sk versus R/k (Figure S-10). The relationship is unusual; S increases with R as long as we keep our markers within the circle. That maybe a testable feature of this case of the unisex circle model.

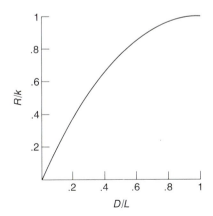

FIGURE S-9
The probability of recombination per sex circle versus the distance between markers as the fraction of the length of the sex circle.

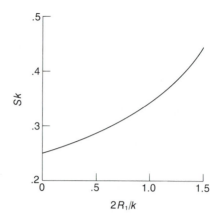

FIGURE S-10
The coefficient of coincidence as a function of recombination frequency for markers within a sex circle.

11-6 (a) The right one will be converted at a slightly higher rate; (b) The right one will be converted at a much higher rate; (c) $\frac{1}{2}$; (d) $\frac{1}{2}$; (e) $\frac{1}{2}$; (f) all; (g) all.

11-7 (a) Let L be the distance from f_L to f_R and x be the position of the marker measured from f_L to f_R. For Holliday's model, l is the splatch length, and the conversion probability is $C = k[e^{-x/l} + e^{-(L-x)/l}]$. For the unisex circle model let L be the distance between the two ends of the circle. Then $C = k(x/L)(L - x)/L$. Note that C for Holliday's model is minimal at $x = L/2$, while C is a maximum at $L/2$ for the unisex circle model.

  (b) On Holliday's model the two most frequent conversion tetrad types containing a $b_1{}^+b_2{}^+$ spore are

$$
\begin{array}{l}
a\ b_1\ b_2{}^+c \\
a\ b_1{}^+b_2{}^+c \\
a^+b_1{}^+b_2\ c^+ \\
a^+b_1{}^+b_2\ c^+
\end{array}
\quad \text{and} \quad
\begin{array}{l}
a\ b_1\ b_2{}^+c \\
a^+b_1{}^+b_2{}^+c \\
a\ b_1{}^+b_2\ c^+ \\
a^+b_1{}^+b_2\ c^+
\end{array}
$$

On the unisex circle model the two most frequent conversion types containing a $b_1{}^+b_2{}^+$ spore are

$$
\begin{array}{l}
a\ b_1\ b_2{}^+c \\
a\ b_1\ b_2{}^+c \\
a^+b_1{}^+b_2{}^+c^+ \\
a^+b_1{}^+b_2\ c^+
\end{array}
\quad
\begin{array}{l}
a\ b_1\ b_2{}^+c \\
a\ b_1\ b_2{}^+c^+ \\
a^+b_1{}^+b_2{}^+c \\
a^+b_1{}^+b_2\ c^+
\end{array}
$$

11-10 We shall consider two cases and conclude that the author's contention is not always valid and may never be.

(a) Case I. Let the splatch be fully asymmetric, which is appropriate for yeast. Let excision tracts be of fixed length and the probability of correction be unity. These assumptions lead to map expansion and rarity of postmeiotic segregation, both of which are reported for yeast. Let the two markers being monitored for single-site conversion and co-conversion be closer together than the length of the excision tract. If there are no other markers nearby, the monitored markers will co-correct whenever they are both in the splatch. Single-site correction will occur when only one of the markers is in the splatch; if correction is on the recipient chain, single-site conversion will result. We diagram the routes to single-site conversion and to co-conversion:

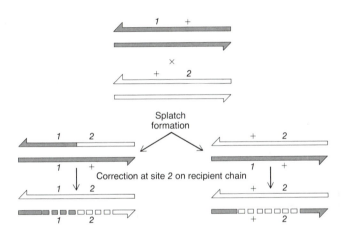

Now add a marker (to the right):

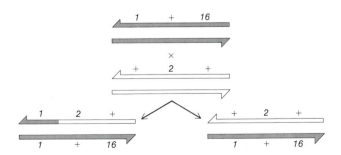

The heteroduplex on the left can give only single-site conversion (for site *2*) whether correction begins at site *2* or *16*; adding the marker at site *16* makes no difference to this class. The chromatid on the right, however, may gain a new fate in the presence of a mismatch at site *16*. If the distance from *16* to *1* is less than the fixed-length excision tract, then site *2* can be corrected separately from site *1* by a tract initiated at *16*:

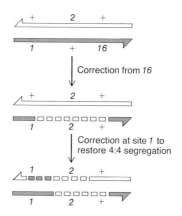

Thus, adding a marker at site *16* has increased single-site conversions (at site *2*) at the expense of co-conversions at sites *1* and *2*. For this case the author's contention is pretty obviously wrong.

(b) Case II. We again suppose correction to be universal so that postmeiotic segregation is rare, and we will suppose splatch formation to be fully asymmetric. We relax the requirement of fixed-length excision tracts, but we suppose splatches to be long so that single-site conversion is always the result of a short tract on a doubly heterozygous splatch. Our starting heteroduplex is like this:

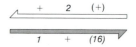

Consider first the cross done in the absence of a marker at site *16*. Let p be the probability that a correction initiated at one site fails to reach the other. The probabilities of the various outcomes of mismatch correction acting on the splatch are as follows:

Normal segregation	$(2-p)/4$
Co-conversion	$(2-p)/4$
Conversion at site *1* only	$p/4$
Conversion at site *2* only	$p/4$

and the ratio of single-site conversions to co-conversions is $2p/(2 - p)$.

Now add the marker at site *16*. What influence can it have on the ratio of single-site to co-conversions? Site *16* can affect this ratio only when it corrects before either site *1* or *2* and only when an excision tract initiated at *16* reaches at least as far as site *2*. If a tract initiated at *16* and having crossed site *2* has a probability $q < p$ of stopping in the *1-2* interval, co-conversions will be increased due to that tract. If the probability q is greater than p, single-site conversions will be increased relative to co-conversions. If $q = p$, the ratio will be unchanged.

When $q = p$, tract lengths are exponentially distributed (see the appendix). When $q > p$, a quasi-fixed length distribution, such as a Gaussian, is implied. In order for q to be less than p, one must imagine a tract-generating mechanism that gets stronger and stronger the further it has traveled. That is a very special and rather weird assumption.

It may be possible to specify a reasonable scheme whereby correction of a nearby site decreases single-site conversion while maintaining or increasing co-conversion, but such is not obvious to me. Try to make one.

APPENDIX

A-1 (a) When $D < l$, mismatch correction plays no role in determining recombination frequency (when l is of fixed length). Following Equation A-5a, we write $R_{l>D<L} = kL(D/L)$. When $D > L$, mismatch correction again plays no role in determining recombination frequency. Following Equation A-5b, evaluated at $a = \frac{1}{2}$, we write $R_{D>L} = kL(1 + D/L)/2$. When $l < D < L$, we can use the third part of Equation A-9:

$$R_{l<D<L} = kL\left[\frac{D}{L} + \frac{\rho(2 - \rho)}{2}\left(1 - \frac{D}{L}\right)\right]$$

This function is graphed in Figure S-11 at $l = 0.3$, and at $\rho = 0.4$ and $\rho = 1$, respectively. The graph for $\rho = 0.4$ is instructive because the inflection at $D/L = 1$ reminds us that LNI has two origins in this model: mismatch correction at $D/L < 1$ and patches at $D/L > 1$.

(b) The dashed line in Figure S-11 corresponds to $l = 0$ and the dotted line to $l > L$. In the former case the mapping function fails to extrapolate

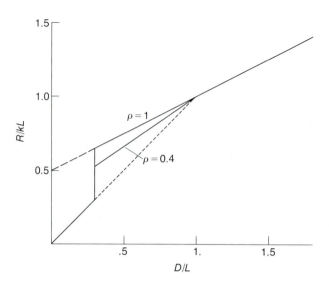

FIGURE S-11
Mapping functions for excision tracts of fixed length (*l*) operating
on splatches of fixed length (*L*), with splices and patches equally
frequent. ρ is the probability of mismatch correction. The upper
limb of the curve extrapolates back to $R/kL > 0$ when the length
of excision tract, *l*, is zero (*dashed line*). The dotted line
corresponds to $\rho = 0$.

to zero; in the latter case the mapping function looks like that for splatches
of fixed length with no excision at all (Figure A-6 at $a = 0.5$).

(c) In Figure S-12 I have plotted SkL versus $2R_1/kL$ (solid line) for the case
of $l = 0.3$ and $\rho = 1$ (which is less tedious to calculate and not much
different from the case for $\rho = 0.4$). The broken line is for SkL for the
fixed-length splatch model (patches and splices equally frequent) with no
correction (taken from Figure A-6 at $a = 0.5$). The discrepancies between
the two models demonstrate the contribution of correction to LNI
($SkL > 0$) and the map expansion resulting from the assumption of fixed-
length excision tracts ($SkL < 0$).

A-2 (a) Equation A-8 gives R for the case where all splatches are splices. We can
weight that expression by a factor of $\frac{1}{2}$ and add to it a term for the contribution made by patches. This term will be composed of two parts: (1) a
contribution due to patches, one end of which falls between the markers;
and (2) a contribution due to mismatch correction operating on patches
that extend across the two marked sites. We can take the first part as
Equation A-6 evaluated at $a = 0$. For the second part we recognize that
the probability of a patch covering both sites is $kLe^{-D/L}$ and that these

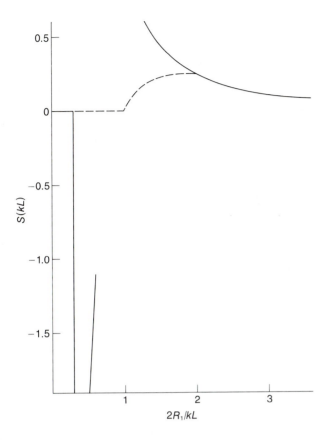

FIGURE S-12
Coefficient of coincidence as a function of recombination frequency for the mapping function in Figure S-11 for $l/L = 0.3$ and $\rho = 1$. The broken line corresponds to $\rho = 0$.

patches will be operated on by mismatch correction in the same way that splices are. Thus, the contribution to R due to patches is

$$kL\left[\left(1 - e^{-\frac{D}{L}}\right) + \frac{\rho(2 - \rho)}{2}e^{-\frac{D}{L}}\left(1 - e^{-\frac{D}{L}/\frac{l}{L}}\right)\right]$$

When we weight patches by $\frac{1}{2}$ and add the term for splices we get

$$R = \frac{kL}{2}\left[\frac{D}{L} + \left(1 - e^{-\frac{D}{L}}\right) + \rho(2 - \rho)e^{-\frac{D}{L}}\left(1 - e^{-\frac{D}{L}/\frac{l}{L}}\right)\right]$$

The first two terms in the bracket describe respectively splices and patches that cover one site but not the other. The last term is the probability that a splatch covers both sites and gets corrected so as to yield a recombinant. To facilitate comparison with previous examples, we again set $l/L = 0.3$ and $\rho = 1$ when we graph R/kL in Figure S-13 (solid line).

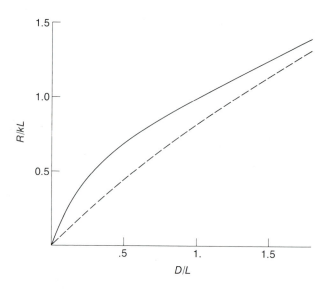

FIGURE S-13
The mapping function for excision tracts of exponential length distribution operating on splatches of exponential length distribution with patches and splices equally frequent. The mean length of tract to splatch (l/L) was set at 0.3; ρ, the probability of correction, was 1.0. The dashed curve is for $\rho = 0$.

FIGURE S-14
The coefficient of coincidence versus recombination frequency for the mapping functions in Figure S-13.

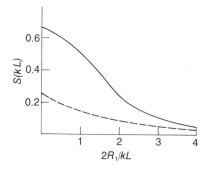

The dashed line is the mapping function for our exponential splatch model with patches and splices equally frequent and no mismatch correction operating (Equation A-6 at $a = 0.5$).

(b) In Figure S-14 the solid line gives SkL for the model in part a, evaluated at $l/L = 0.3$ and $\rho = 1$. The dashed line is SkL for the exponential splatch model ($a = 0.5$) with no correction.

A-4 Even if you cannot find the flying factor of two, you should at least recognize that there is no way to get map expansion without involving marker effects, and the explanation quoted does not do so.

Glossary

The terms defined here are not specific to recombination studies and should be well known to most readers. Definitions for terms defined in the text can be found by referring to the index.

Ascus.   A fungal spore sac.

Auxotroph.   An individual or strain with a genetically imposed nutritional requirement beyond those of the wild type, or prototrophic, strain of origin.

Base.   Adenine (A), guanine (G), thymine (T), and cytosine (C).

Base pair.   In a DNA duplex, A on one chain is paired with T on the other; G on one chain is paired with C on the other.

Centromere.   Spindle fiber attachment point on a eukaryotic chromosome.

Chromatin.   The material (DNA and protein) of which chromosomes are composed.

Cistron.   A gene, operationally defined as a unit of complementation.

Codon.   A triplet of adjacent nucleotides in a gene coding for a single amino acid in the protein product of that gene.

Concatemer.   A long DNA molecule made up of a linearly repeated chromosome.

Conjugation.   In bacteria the temporary union of male (F^+) and female (F^-) cells that allows the transfer of a chromosome or part thereof from the F^+ to the F^- cell.

Cotranscribed.   The condition in which two nearby genes are copied onto a single mRNA molecule.

Diploid.   Containing two homologous sets of chromosomes.

Encapsidation. The process of covering a viral chromosome with a protein coat or capsid.

Episome. A multigenic piece of DNA that can exist either as part of a bacterial chromosome or as an entity replicating in a bacterial cell unconnected with the chromosome.

Eukaryote. Organisms whose cells contain true nuclei, characterized by a mitotic cycle in which elongated chromosomes are replicated during an interphase, condensed in prophase, oriented on a spindle in metaphase, and distributed to daughter cells in anaphase. The chromosomes in eukaryotes are separated from the cytoplasm by a porous nuclear membrane except during metaphase and anaphase.

Frameshift mutations. Deletions and additions of small numbers of base pairs that put the translation apparatus out of reading frame.

Gamete. A germ cell whose union with another gamete initiates the diplophase.

Gene. Segment of a chromosome providing information for the conduct of a particular task, usually that of dictating the amino acid sequence of a protein.

Genotype. A statement of the relevant informational content of a cell; a list of the mutant genes of interest in a cell.

Haploid. Containing one set of chromosomes.

Haplophase. Haploid stage in a life cycle.

Heterothallic. Referring to fungi having two distinct mating types or sexes in haplophase.

Heterozygote. In eukaryotes, a diploid cell resulting from the union of genetically different gametes. In phages, a particle that contains two alleles of a single gene by virtue either of the duplex structure of DNA or of a terminal redundancy.

Homologous. Related by descent from a common ancestor. Specifically, two chromosomes are homologous when they have approximately the same nucleotide sequence.

Homothallic. Referring to fungi that do not have distinct mating types or sexes in haplophase.

Lysogenization. The act whereby infecting phages transform a cell to the state of lysogeny.

Lysogeny. The hereditary harboring of a phage chromosome.

Lytic growth. Phage development culminating in cell lysis with release of phage particles.

Meiocyte. A cell undergoing meiosis.

Meiosis. The process of chromosome distribution during the two sequential cell divisions that reduce a diploid cell (bearing chromosomes whose DNA has replicated) to a set of four haploid cells.

Mitosis. The process of chromosome distribution into daughter cells that assures that each daughter receives the numbers and types of chromosomes that characterized the parent cell.

Mutation. A change in the nucleotide sequence of a chromosome. Generally applied to those changes that have consequences for the phenotype.

Okazaki piece. A short, newly synthesized piece of DNA chain found near a replicating fork.

Oligonucleotide. A rather short polynucleotide chain.

Phenotype. A statement of relevant aspects of the appearance of an organism.

Polynucleotide. A polymer of nucleotides connected in linear array. DNA in the Watson-Crick structure is a pair of structurally complementary polynucleotides wound about each other in a "double helix."

Repressor. A protein that binds to a specific site (or sites) on a chromosome, thereby preventing transcription of adjacent genes.

Telomere. The (stable) end of a eukaryotic chromosome.

Transcription. The process of making a RNA chain along a chromosomal, DNA template.

Transduction (phage-mediated). The transfer of genes from one (bacterial) cell to another within a phage capsid.

Transfection. Infection of a cell by viral nucleic acid in the absence of the usual viral protein capsid.

Translation. The process of making a polypeptide chain along an RNA template.

Virion. A complete virus particle composed of a nuclei acid chromosome surrounded by a protein coat (capsid).

Journal Abbreviations

AG	Advances in Genetics
AIP	Annales de L'Institute Pasteur
AJB	American Journal of Botany
AJBS	Australian Journal of Biological Sciences
ARG	Annual Review of Genetics
ARM	Annual Review of Microbiology
ASNBBV	Annales des Sciences Naturelles Botanique Végétale Biologie
BBRC	Biochemical and Biophysical Research Communications
BJ	Biophysical Journal
C	Chromosoma
CRASP	Comptes Rendus Hebdomadaires des Séances de L' Académie des Sciences, Paris
CRTCL	Comptes Rendus des Travaux Carlsberg Lab
CSHSQB	Cold Spring Harbor Symposia of Quantitative Biology
G	Genetics
GR	Genetical Research
H	Heredity
JB	Journal of Bacteriology
JBC	Journal of Biological Chemistry
JCB	Journal of Cell Biology
JCP	Journal of Cellular Physiology
JChPh	Journal de Chimie Physique
JEZ	Journal of Experimental Zoology

JG	Journal of Genetics
JGP	Journal of General Physiology
JMB	Journal of Molecular Biology
JTB	Journal of Theoretical Biology
JV	Journal of Virology
MGG	Molecular and General Genetics
MR	Mutation Research
N	Nature
NNB	Nature New Biology
PNAS	Proceedings of the National Academy of Sciences, USA
RR	Radiation Research
S	Science
V	Virology
ZN	Zeitschrift für Naturforschung
ZV	Zeitschrift für Vererbungslehre

Index